Arduino BLINK Blueprints

Get the most out of your Arduino to develop exciting and creative LED-based projects

Samarth Shah

Utsav Shah

[PACKT] open source*
PUBLISHING   community experience distilled

BIRMINGHAM - MUMBAI

Arduino BLINK Blueprints

First published: May 2016

Production reference: 1240516

Published by Packt Publishing Ltd.
Livery Place
35 Livery Street
Birmingham B3 2PB, UK.

ISBN 978-1-78528-418-2

www.packtpub.com

Credits

About the Authors

Samarth Shah is a software engineer by profession and maker by heart. He leads maker activities at Pune Makers and heads Infosys Robotics Club. He loves building creative/innovative prototypes using the latest hardware/sensors (Raspberry Pi, Arduino, Kinect, Leap Motion, and many more) and software. He has given talks at various national and international conferences. He has authored a book on Raspberry Pi entitled *Learning Raspberry Pi, Packt Publishing*. During the day, he works on various data visualization techniques and UI frameworks. At night, he does blogging, reading, writing, and many more things. You can read more about him at http://samarthshah.com.

Utsav Shah is an instrumentation engineer who loves to work on the latest hardware as well as software technologies. He has been featured on India's leading website http://yourstory.in and Ahmedabad Mirror (Times Group) for his research work on "Converting sign language into speech" using a Leap Motion controller. Apart from his regular work at Infosys Limited, he manages activities of Infosys Robotics Club. In his leisure time, he loves to read books and work on cutting-edge technologies.

We would like to thank our parents Pareshbhai and Sandhyaben for their constant encouragement and continuous support.

About the Reviewer

Timothy Gorbunov was born in the USA. At a young age, he fell in love with building and constructing things just like his dad. Tim became very good at Origami and started to sell it at elementary school. As he grew up, Tim leaned more towards electronics because it fascinated him more than any other hobby. Creating circuits that buzzed or flashed was one of Tim's favorite things to do. As time went by, he started exploring more advanced electronics and programming, and from that point on, he became more and more knowledgeable in electronics. He got hired to help create cymatic light shows at Cymaspace. There, he helped start Audiolux devices, a company that specializes in sound reactive technologies, by helping design their products. Tim does many other things other than electronics, such as fishing and hiking, but most importantly Tim believes in God. Tim spends a lot of time studying the Bible, praying, and going to church. He wants everyone to find the truth, the fact that Jesus Christ died for every person here on earth to bring redemption from their sins and give everlasting life with him in heaven. Tim also reviewed another book, *Arduino by Example*, *Packt Publishing*, in which he gained valuable experience in reviewing.

> I thank opportunities such as this one for which I can use what I know to help bring books to life. I thank many people in my life who have allowed me to start my interest in this field as well as the Internet for being such a great resource for answering many of questions.

www.PacktPub.com

eBooks, discount offers, and more

Did you know that Packt offers eBook versions of every book published, with PDF and ePub files available? You can upgrade to the eBook version at www.PacktPub.com and as a print book customer, you are entitled to a discount on the eBook copy. Get in touch with us at customercare@packtpub.com for more details.

At www.PacktPub.com, you can also read a collection of free technical articles, sign up for a range of free newsletters and receive exclusive discounts and offers on Packt books and eBooks.

https://www2.packtpub.com/books/subscription/packtlib

Do you need instant solutions to your IT questions? PacktLib is Packt's online digital book library. Here, you can search, access, and read Packt's entire library of books.

Why subscribe?

- Fully searchable across every book published by Packt
- Copy and paste, print, and bookmark content
- On demand and accessible via a web browser

Table of Contents

Preface

Arduino is an open source prototyping platform based on easy-to-use hardware and software. Arduino has been used in thousands of different projects and applications by a wide range of programmers and artists, and their contributions have added up to an incredible amount of accessible knowledge that can be of great help to novices and experts alike.

This book will be your companion to bring out the creative genius in you. As you progress through the book, you will learn how to develop various projects with Arduino.

What this book covers

Chapter 1, *Getting Started with Arduino and LEDs*, introduces you to different Arduino boards followed by installation instructions for the Arduino IDE. You will write a "Hello World" program using the Arduino IDE and will learn about serial communication.

Chapter 2, *Project 1 – LED Night Lamp*, presents you with some cool stuff of controlling LEDs and will show you how to control different LEDs with an artistic approach.

Chapter 3, *Project 2 – Remote Controlled TV Backlight*, teaches you the basics of IR LEDs and the basics of IR communication. Once you have learnt about programming IR sensor, you will use an IR sensor to control the TV backlight using a remote.

Chapter 4, *Project 3 – LED Cube*, introduces you to soldering in detail. You will also understand how to create a 4*4*4 LED Cube using the Arduino UNO board.

Chapter 5, *Sound Visualization and LED Christmas Tree*, shows you how to visualize sound using Arduino and then we will develop an LED Christmas tree.

Chapter 6, Persistence of Vision, helps us create an even more intensive experience by moving LEDs using motors. You will create a Persistence of Vision wand using an LED array and a motor.

Chapter 7, Troubleshooting and Advanced Resources, starts with common troubleshooting techniques. The second and last part of the chapter discusses resources that will be useful if you want to do advanced stuff with Arduino.

What you need for this book

All you need is an Arduino IDE and the enthusiasm to work on interesting projects.

Who this book is for

Anyone with basic computer knowledge should be able to get the most out of this book. Although familiarity with some of the electronics would be helpful, it is not a must.

Conventions

In this book, you will find a number of text styles that distinguish between different kinds of information. Here are some examples of these styles and an explanation of their meaning.

Code words in text, database table names, folder names, filenames, file extensions, pathnames, dummy URLs, user input, and Twitter handles are shown as follows: "If you have downloaded "Windows ZIP file for non admin install", extract it and you will find arduino.exe."

A block of code is set as follows:

```
// the setup function runs once when you press reset or power the
board
void setup() {
  // initialize digital pin 13 as an output.
  pinMode(13, OUTPUT);
}
```

New terms and **important words** are shown in bold. Words that you see on the screen, for example, in menus or dialog boxes, appear in the text like this: "Make sure you have selected **Arduino UNO** under the **Tools | Board** section."

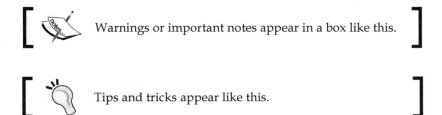

[Warnings or important notes appear in a box like this.]

[Tips and tricks appear like this.]

Reader feedback

Feedback from our readers is always welcome. Let us know what you think about this book—what you liked or disliked. Reader feedback is important for us as it helps us develop titles that you will really get the most out of.

To send us general feedback, simply e-mail feedback@packtpub.com, and mention the book's title in the subject of your message.

If there is a topic that you have expertise in and you are interested in either writing or contributing to a book, see our author guide at www.packtpub.com/authors.

Customer support

Now that you are the proud owner of a Packt book, we have a number of things to help you to get the most from your purchase.

Downloading the example code

You can download the example code files for this book from your account at http://www.packtpub.com. If you purchased this book elsewhere, you can visit http://www.packtpub.com/support and register to have the files e-mailed directly to you.

You can download the code files by following these steps:

1. Log in or register to our website using your e-mail address and password.
2. Hover the mouse pointer on the **SUPPORT** tab at the top.
3. Click on **Code Downloads & Errata**.
4. Enter the name of the book in the **Search** box.
5. Select the book for which you're looking to download the code files.
6. Choose from the drop-down menu where you purchased this book from.
7. Click on **Code Download**.

You can also download the code files by clicking on the **Code Files** button on the book's webpage at the Packt Publishing website. This page can be accessed by entering the book's name in the **Search** box. Please note that you need to be logged in to your Packt account.

Once the file is downloaded, please make sure that you unzip or extract the folder using the latest version of:

- WinRAR / 7-Zip for Windows
- Zipeg / iZip / UnRarX for Mac
- 7-Zip / PeaZip for Linux

The code bundle for the book is also hosted on GitHub at `https://github.com/PacktPublishing/Arduino-BLINK-Blueprints`. We also have other code bundles from our rich catalog of books and videos available at `https://github.com/PacktPublishing/`. Check them out!

Downloading the color images of this book

We also provide you with a PDF file that has color images of the screenshots/diagrams used in this book. The color images will help you better understand the changes in the output. You can download this file from `https://www.packtpub.com/sites/default/files/downloads/ArduinoBLINKBlueprints_ColorImages.pdf`.

Errata

Although we have taken every care to ensure the accuracy of our content, mistakes do happen. If you find a mistake in one of our books — maybe a mistake in the text or the code — we would be grateful if you could report this to us. By doing so, you can save other readers from frustration and help us improve subsequent versions of this book. If you find any errata, please report them by visiting `http://www.packtpub.com/submit-errata`, selecting your book, clicking on the **Errata Submission Form** link, and entering the details of your errata. Once your errata are verified, your submission will be accepted and the errata will be uploaded to our website or added to any list of existing errata under the Errata section of that title.

To view the previously submitted errata, go to `https://www.packtpub.com/books/content/support` and enter the name of the book in the search field. The required information will appear under the **Errata** section.

Piracy

Piracy of copyrighted material on the Internet is an ongoing problem across all media. At Packt, we take the protection of our copyright and licenses very seriously. If you come across any illegal copies of our works in any form on the Internet, please provide us with the location address or website name immediately so that we can pursue a remedy.

Please contact us at copyright@packtpub.com with a link to the suspected pirated material.

We appreciate your help in protecting our authors and our ability to bring you valuable content.

Questions

If you have a problem with any aspect of this book, you can contact us at questions@packtpub.com, and we will do our best to address the problem.

1
Getting Started with Arduino and LEDs

Welcome to the exciting world of physical computing! Today, hobbyists and experts all over the world use Arduino to make interactive objects and to create cool prototypes. In this chapter, you will get introduced to different Arduino boards, followed by installation instructions for Arduino IDE. You will write a "Hello World" program using Arduino IDE and will learn about serial communication. By the end of this chapter, you will have a basic knowledge of Arduino and its IDE, which will be helpful in the remaining chapters of the book. In this chapter, we will cover the following:

- Arduino boards
- Arduino IDE
- Before you start
- "Hello World"
- Using serial communication
- The world of LED

Arduino boards

Arduino was originally created for artists and designers as an easy and quick prototyping tool. Designers were able to create sophisticated designs and artworks even without having knowledge of electronics and programming. So it is understood that the first few steps of learning Arduino are very easy. In this section, you will get introduced to different Arduino boards and learn how to choose an Arduino board for your project and some information on the Arduino UNO board, which you will be using throughout this book.

Different Arduino boards

Beginners often get confused when they discover Arduino projects. When looking for Arduino, they hear and read such terms as Uno, Zero, and Lilypad. The thing is, there is no such thing as "Arduino". In 2006, the Arduino team designed and developed a microcontroller board and released it under an open source license. Over the years, the team has improved upon the design and released several versions of the boards. These versions mostly had Italian names. There are numbers of boards that the team has designed over the past 10 years:

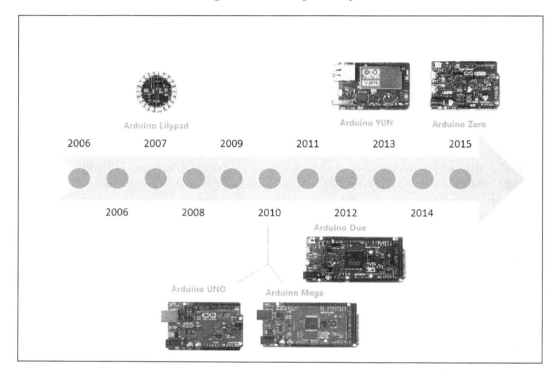

The Arduino team didn't just improve on design, they invented new designs for specific use cases in mind. For example, they developed Arduino Lilypad to embed a board into textiles. It can be used to build interactive T-shirts. The following table shows the capability of different Arduino boards. Arduino boards may differ in their appearance, but they have a lot in common. You can use the same tools and libraries to program:

Name	Processor	Dimension	Voltage	Flash(kB)	Digital I/O(PWM) pins	Analog input pins
Arduino Lilypad	ATmega168V	51 mm outer diameter	2.7-5.5 V	16	14(6)	6
Arduino YUN	Atmega32U4	68.6 mm × 53.3 mm	5 V	32	14(6)	12
Arduino Mega	ATmega2560	101.6 mm × 53.3 mm	5 V	256	54(15)	16
Arduino Due	ATSAM3X8E	101.6 mm × 53.3 mm	3.3 V	512	54(12)	12
Arduino Zero	ATSAMD21G18A	68.6 mm × 53.3 mm	3.3 V	256	14(12)	6
Arduino UNO	ATmega328P	68.6 mm × 53.3 mm	5 V	32	14(6)	6

As Arduino boards' design and schematics are open source, anyone can use and change the original board design and can create their own version of an Arduino-compatible board. Because of that, you can find countless Arduino clones on the web which can be programmed using the same tools and libraries used for original Arduino boards.

How to choose an Arduino board for your project

With so many options available, it becomes challenging for a person to decide which Arduino board to use for a project. The Arduino family is huge, and it is impossible to read about each and every board and decide upon which board to use for a particular project.

The following flowchart simplifies the process by providing a decision tree for widely-used Arduino boards and the most common use cases/applications:

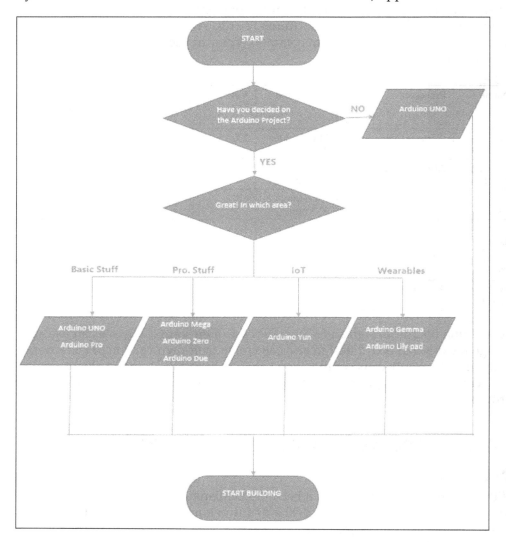

If you are not sure what you will build and what hardware capabilities are required, start building your prototype using Arduino UNO. Arduino UNO has the best documentation and best support. It is also the most compatible of all Arduino boards. Most of the existing libraries and shields are compatible with Arduino UNO. And finally, most of the code that has been written on earlier versions of Arduino boards will also work with Arduino UNO.

Arduino shields are modular circuit boards that can be plugged on top of the Arduino PCB, extending its capabilities. Want to connect your Arduino to the Internet? There is a shield for that. There are hundreds of shields available online, which makes Arduino more than just a development board.

Throughout this book, we will use the Arduino UNO board.

Arduino UNO

"UNO" means one in Italian, and it was chosen to mark the release of Arduino IDE v1.0. It is the first in a series of USB Arduino boards. As mentioned before, it is one of the most widely used boards in the Arduino Family:

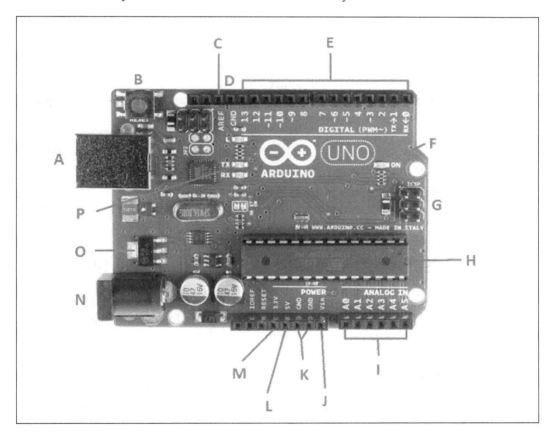

In this section you will get introduced to different components of the Arduino UNO board.

- **A**: **USB plug**: Every Arduino board needs a way to be connected to a power source. The Arduino UNO can be powered from a USB cable coming from your computer. The USB connection is also how you will load code onto your Arduino board.

- **B**: **Reset button**: Pushing it will temporarily connect the reset pin to ground and restart any code that is loaded on the Arduino board.

- **C**: **AREF**: This stands for Analog Reference. In this book, you are not going to use this pin. It is sometimes used to set an external reference voltage (between 0 and 5) as the upper limit for the analog input pins.

- **D**: **GND**: Digital ground.

- **E**: **Pin 0 to Pin 13**: The area of pins under the **DIGITAL** label are digital pins. These pins can be used for both digital input (like telling if a button is pushed) and digital output (like powering an LED). Next to some of the pins (3, 5, 6, 9, 10, and 11) there is a tilde (~) sign, which means those pins can also act as pulse width modulation apart from normal digital pins. PWM (pulse width modulation) is a technique for getting an analogue signal with digital means by controlling on and off duration of the signal.

- **F**: **ON**: This is a power LED indicator. This LED should light up whenever you plug your Arduino into a power source. If the LED doesn't turn ON, there is something wrong with your Arduino board.

- **G**: **In-circuit serial programmer**: You will use these pins very rarely. Mostly, hardware manufacturers use these pins for debugging purposes. Also, these pins are used to program Arduino using other Arduino or other microcontrollers.

- **H**: **Main IC**: This is an ATmega 328 microcontroller. Think of it as the brain of your Arduino board.

- **I**: **Pin A0 to Pin A5**: The area of pins under the **Analog In** label are Analog In pins. These pins can read the signal from an analog sensor (like a temperature sensor) and convert it into a digital value that can be read by Arduino software and can be used for further processing.

- **J**: **Vin**: Voltage In. Arduino can be supplied with power from a DC jack, the USB connector, or Vin pin. While supplying with Vin, you can give up to 12 V. The in-built regulator in Arduino will take care of regulating voltage to 5 V.

- **K**: **GND**: Ground pin.

- **L: 5 V**: Power pin which supplies 5 volts.
- **M: 3.3 V**: Power pin which supplies 3.3 volts.
- **N: External power supply**: Once you have uploaded your code to Arduino you don't need a computer just to draw power. You can use an external power supply and can use Arduino as a standalone device.
- **O: Voltage regulator**: The voltage regulator is not actually something you can interact with on Arduino. It controls the amount of voltage that is let into the Arduino board.
- **P: Tx Rx LEDs**: Tx is short for transmit, RX is short for receive. In electronics, TxRX are used to indicate the pins responsible for serial communication. These LEDs will give nice visual indications whenever our Arduino is receiving or transmitting data.

Arduino IDE

As, initially, Arduino was initially designed for artists and designers, the Arduino team has tried to develop Arduino software (IDE) as simply as possible without compromising on the power of the tool. Before you run your "Hello World" program, you need to install Arduino IDE on your computer.

Installing Arduino IDE

Arduino IDE is supported on all major operating systems, initially Windows, Mac, and Linux.

On Windows

1. Go to `https://www.arduino.cc/en/Main/Software`.
2. Download "Windows Installer" or "Windows ZIP file for non admin install".
3. If you have downloaded "Windows Installer", double click on it and it will be installed.
4. If you have downloaded "Windows ZIP file for non admin install", extract it and you will find `arduino.exe`. Double click on it to get started with the Arduino IDE.

On Linux

1. Go to `https://www.arduino.cc/en/Main/Software`.

2. Download "Linux 32 bits" or "Linux 64 bits" depending on your OS type.

3. Extract it and run the Arduino executables to get started with the Arduino IDE.

On Mac

1. Go to `https://www.arduino.cc/en/Main/Software`.

2. Download "Mac OS X 10.7 Lion or newer".

3. Extract it and run the Arduino executables to get started with the Arduino IDE.

Understanding Arduino IDE

If you have used IDEs such as Visual Studio, XCode, and Eclipse before, you'd better lower your expectations, because Arduino IDE is very simple. It mainly consists of an editor, a compiler, a loader, and a serial monitor. It has no advanced features such as a debugger or code completion:

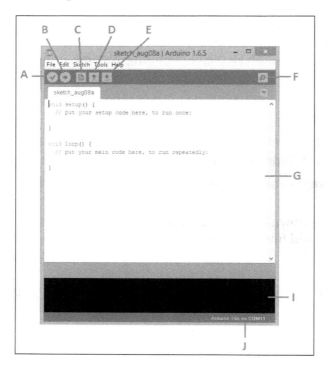

Let's look into each button separately:

- **A: Verify button**: This will compile the program that's currently in the editor. It does not only verify the program syntactically. It also turns it into a representation suitable for the Arduino board.

- **B: Upload button**: This will compile and upload the current program to the connected Arduino Board.

- **C: New button**: This creates a new program by opening a new editor window.

- **D: Open button**: With this button, you can open an existing program from the file system.

- **E: Save button**: This saves the current program.

- **F: Serial monitor**: Arduino can communicate with a computer via a serial connection. Clicking the Serial Monitor button opens a serial monitor window that allows you to watch the data sent by Arduino and also to send data back. You will learn more about it in the *Using serial communication* section.

- **G: Editor window**: This is where you will write the code.

- **H: Error console**: This is where you will see all the error messages of your code.

- **I: Status bar**: This will show the connected Arduino board name along with the COM port number.

Before you start

After getting some understanding of Arduino UNO and IDE, there are a couple of things that you need to know before you dive into the world of Arduino.

Power supply

As mentioned in the *Arduino UNO* section, there are two ways you can power up your Arduino UNO. One is by using a USB cable connected to your computer and the second one is by a 12 V external power supply. Please make sure that you don't use a power supply greater than 20 volts as you will overpower and thereby destroy your Arduino Board. The recommended voltage for most Arduino models is between 6 and 12 volts.

Verifying connection

This is the last step before you write your "Hello World" program with your Arduino:

1. Make sure you have selected **Arduino UNO** under the **Tools | Board** section. If you have some other Arduino board, make sure you select that board:

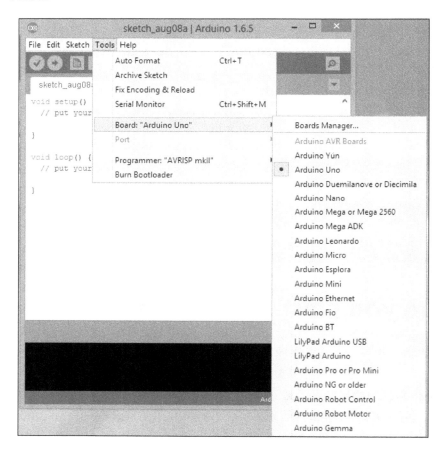

2. Select the COM port from **Tools | Port**, to which your Arduino UNO board is connected. In the following image it is COM13, but it will vary from computer to computer:

"Hello World"

You must be aware of "Hello World" programs from computer science, where you write a piece of code which will display "Hello World". In electronics hardware board space, "Hello World" refers to blinking an LED by writing a simple program.

The resistor will block the flow of current in both directions. A diode is a two-terminal electronics component that has low resistance in one direction and high resistance in the other direction. Diodes are mostly made up of silicon. LEDs are the most commonly used diodes in any electronics circuit. LED stands for light emitting diode, so it emits light when sufficient voltage is provided across the LED anode and cathode.

The longer lead of the LED is the anode and the other end is the cathode. The color of the light depends on the semiconductor material used in the LED, as shown in the following diagram:

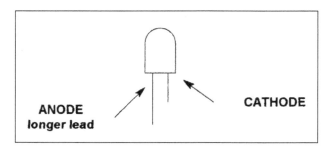

Connect the longer lead of the LED to pin 13 and shorter lead of the LED to GND (ground pin) and write the following code in a new editor window:

```
// the setup function runs once when you press reset or power the
board
void setup() {
  // initialize digital pin 13 as an output.
  pinMode(13, OUTPUT);
}

// the loop function runs over and over again forever
void loop() {
  digitalWrite(13, HIGH);    // turn the LED on (HIGH is the voltage
level)
  delay(1000);               // wait for a second(1000 millisecond)
  digitalWrite(13, LOW);     // turn the LED off by making the voltage
LOW
  delay(1000);               // wait for a second(1000 millisecond)
}
```

You will notice from the preceding code that there are two important functions in each program. The first one is setup, which runs only once when you power up the board or press the reset button. The second function is loop, which runs over and over forever.

In the setup function, you should write a code that needs to be executed once, like defining a variable, initializing a port as INPUT or OUTPUT. In the preceding code, digital pin 13 is defined as OUTPUT. In the loop function, the first line will put the HIGH voltage on pin 13, which will turn on the LED connected to pin 13.

The "Hello World" program will turn the LED on for one second and turn off the LED for one second. The `delay(1000)` function will induce a delay of one second and after that, `digitalWrite(13, LOW)` will put the low voltage on pin 13, which will turn off the LED. Again, before you turn the LED on, you need to wait for one second by putting `delay(1000)` at the end of the code.

Using serial communication

Serial communication is used for communication between the Arduino board and a computer or other devices. All Arduino boards have at least one serial port which is also known as a UART. Serial data transfer is when we transfer data one bit at a time, one right after the other. Information is passed back and forth between the computer and Arduino by, essentially, setting a pin to high or low. Just like we used that technique to turn an LED on and off, we can also send data. One side sets the pin and the other reads it.

In this section, you will see two examples. In the first example, Arduino will send data to the computer using serial communication, while in the second example, by sending a command (serial) from the computer, you can control the functionality of the Arduino board.

Serial write

In this example, the Arduino board will communicate with the computer using the serial port, which can be viewed on your machine using the Serial Monitor.

Write the following code to your Arduino editor:

```
void setup()                       // run once, when the sketch starts
{
  Serial.begin(9600);              // set up Serial library at 9600 bps

  Serial.println("Hello world!");  // prints hello with ending line
break
}

void loop()                        // run over and over again
{
                                   // do nothing!
}
```

Even if you have nothing in the setup or loop procedures, Arduino requires them to be there. That way it knows you really mean to do nothing, as opposed to forgetting to include them!

`Serial.begin` sets up Arduino with the transfer rate we want, in this case 9600 bits per second. `Serial.println` sends data from Arduino to the computer.

Once you compile and upload it to your connected Arduino board, open **Serial Monitor** from the Arduino IDE. You should be able to see the **Hello world!** text being sent from the Arduino board:

 If you have trouble locating **Serial Monitor**, check the *Understanding Arduino IDE* section of this chapter.

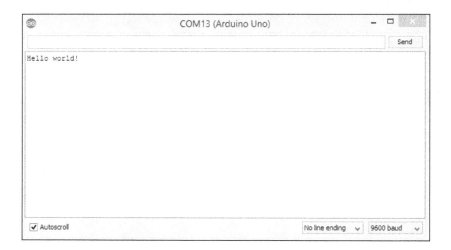

Serial read

In the previous example, serial library was used to send a command from Arduino to your computer. In this example, you will send a command from the computer, and Arduino will do a certain operation (turn on/off LED) based on the command received:

```
int inByte; // Stores incoming command

void setup() {
```

```
Serial.begin(9600);
pinMode(13, OUTPUT); // LED pin
Serial.println("Ready"); // Ready to receive commands
}
void loop() {
  if(Serial.available() > 0) { // A byte is ready to receive
    inByte = Serial.read();
    if(inByte == 'o') { // byte is 'o'
      digitalWrite(13, HIGH);
      Serial.println("LED is ON");
    }
    else
    {
      // byte isn't 'o'
      digitalWrite(13, LOW);
      Serial.println("LED is OFF");
    }
  }
}
```

The inByte function will store the incoming serial byte. From the previous example, you should be familiar with the commands written in the setup function. In the loop function, first you need to know when a byte is available to be read. The Serial. available() function returns the number of bytes that are available to be read. If it is greater than 0, Serial.read() will read the byte and store it in an inByte variable. Let's say you want to turn on the LED when the letter 'o' is available. For that you will be using the if condition, and you will check whether the received byte is 'o' or not. If it is 'o', turn on the LED by setting pin 13 to HIGH. Arduino will also send an **LED is ON** message to the computer, which can be viewed in Serial Monitor:

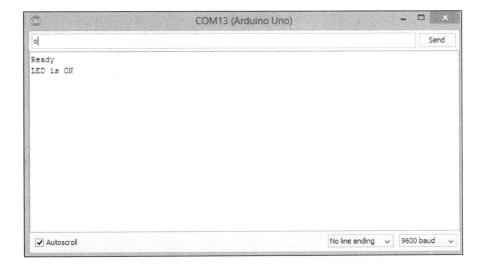

If it is any other character, then turn off the LED by setting pin 13 to LOW. Arduino will also send an **LED is OFF** message to the computer, which can be viewed in Serial Monitor:

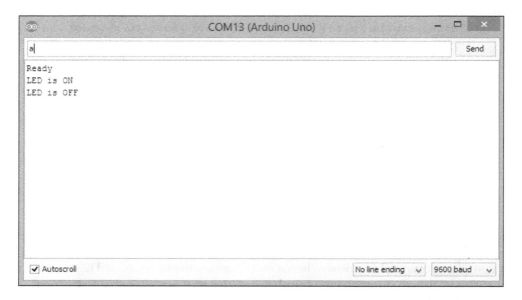

The world of LED

LED stands for light emitting diode, so it emits light when sufficient voltage is provided across the LED anode and cathode. Today's LEDs are available in many different types, shapes, and sizes – a direct result of the tremendous improvements in semiconductor technology over recent years. These advancements have led to better illumination, longer service life, and lower power consumption. They've also led to more difficult decision making, as there are so many types of LED to choose from.

LEDs can be categorized into miniature, high power, and application-specific LEDs:

- **Miniature LEDs**: These LEDs are extremely small and usually available in a single color/shape. They can be used as indicators on devices such as cell phones, calculators, and remote controls.

- **High power LEDs**: Often referred to as high output LEDs, these are a direct result of improved diode technology. They offer a much higher lumen output than standard LEDs. Typically, these LEDs are used in car headlights.

- **Application-specific LEDs**: As the name suggests, there are many LEDs that fall under this category. These are flash LEDs, RGB LEDs, seven segment display, LED lamps, and LED bars.

Summary

In this chapter, an overview of different Arduino boards was covered, with a detailed explanation of an Arduino UNO board. Arduino IDE was explained, with installation instructions for Windows, Mac, and Linux machines. "Hello World" and a serial communication Arduino sketch were developed.

In the next chapter, we will develop an LED mood lamp and you will learn some artistic stuff apart from programming LEDs.

2
Project 1 – LED Night Lamp

In *Chapter 1, Getting Started with Arduino and LEDs,* you learned about the "Hello World" of physical computing. Now, as you have basic knowledge of Arduino and its IDE, we can go ahead with some cool stuff to do with controlling LEDs. The following topics will be covered in this chapter:

- Introduction to breadboard
- Controlling multiple LEDs
- LED fading
- Creating a mood lamp
- Developing an LED night lamp

By the end of this chapter, you will be able to control different LEDs in an artistic approach.

Introduction to breadboard

Prototyping is the process of testing out an idea by creating a preliminary model from which other products can be developed or formed. The breadboard is one of the most fundamental pieces for prototyping electronics circuits. As it does not require any soldering, it is also referred to as a "solderless board".

Structure of a breadboard

Almost all modern breadboards are made up of plastic. A modern breadboard consists of a perforated block of plastic with numerous metal clips under the perforations. The breadboard has strips of metal underneath the board and holes on top of the board. In the following image, you can see the structure of the breadboard:

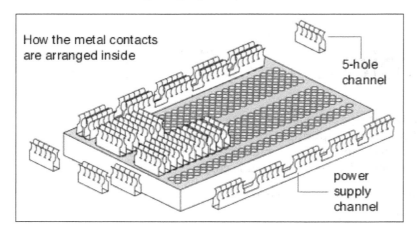

The main structure of the breadboard is made up of a main central area, which is a block of two sets of columns, where each column is made up of many rows. All of these rows are connected on a row-by-row basis.

Using a breadboard

The breadboard has many strips of copper beneath the board that connects the holes as shown (short circuited or same potential). The upper blue lines are not connected with the lower ones. In the case of electronic circuits, a power supply is required at various pins. So instead of making many connections with a power supply, one can give a power supply to one of the holes on the breadboard and can get a power supply from its outer holes:

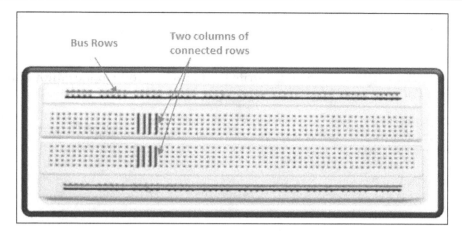

Multiple breadboards can be connected together to form larger prototyping board experimenting areas. If you want to use any chip/IC (integrated circuit), you can place it with one side on the upper block and the other side on the lower block of the breadboard. It will be very easy for us to understand the breadboard, as we will be using the breadboard extensively for all our projects.

Controlling multiple LEDs

In the embedded world, controlling a single LED is "Hello World" code, which we learned in first chapter. Now, as we are familiar with the concepts of LED, we can start to control multiple LEDs with Arduino. Here, we will start by making a simple traffic light module.

Simple traffic light controller

As we all know, a traffic light is made up of three LEDs: red, yellow, and green. To make this project, we need red, green, and yellow LEDs, strip wires, and a few 255 Ω resistors.

In the previous chapter, in our "Hello World" program, we connected an LED directly with pin 13. Here, we will connect red, yellow, and green LEDs with pins 9, 10, and 11 respectively. In the case of pin 13, it has an in-built pull up resistor. Pull up resistors are used to limit the current supplied to an LED. We can't give current of more than a few mA to LEDs. But, with pin 13, current is itself limited in Arduino by the internal pull up resistor. If we want to connect to pins other than 13, we need to add resistors in the circuit.

Connect the longer head of the red LED to pin 9 and the shorter head to the resistor and then the resistor to ground. Make a similar connection with the yellow and green LEDs to pins 10 and 11 respectively.

Once you have made the connection, write the following code in the editor window of Arduino IDE:

```
// Initializing LEDs for using it in the code.
// Initializing outside any function so global variable can be
accessed throughout the code

int redLed = 9;
int yellowLed = 10;
int greenLed = 11;

// the setup function runs once when you press reset or power the
board
void setup() {
  // initialize digital pin for red, yellow and green led as an
output.
  pinMode(redLed, OUTPUT);
  pinMode(yellowLed, OUTPUT);
  pinMode(greenLed, OUTPUT);
}

// the loop function runs over and over again forever
void loop() {
  digitalWrite(redLed, HIGH);      //    Making red led high
  digitalWrite(yellowLed, LOW);    //    Making yellow led low
  digitalWrite(greenLed, LOW);     //    Making green led low
  delay(10000);                    //    Wait for 10 seconds (10000
milliseconds)
  digitalWrite(redLed, LOW);
  //    Making red led low
  digitalWrite(yellowLed, LOW);    //    Making yellow led low
  digitalWrite(greenLed, HIGH);    //    Making green led high
  delay(10000);                    //    Wait for 10 seconds (10000
milliseconds)
  digitalWrite(redLed, LOW);       //    Making red led low
  digitalWrite(yellowLed, HIGH);   //    Making yellow led high
  digitalWrite(greenLed, LOW);     //    Making green led low
  delay(3000);                     //    Wait for 3 seconds 3000
milliseconds)
}
```

In the preceding code, you can see that it is much the same as the "Hello World" program, except here we are controlling multiple LEDs. Here, we are initializing the variables as integers and using that same variable throughout the code. So, in the future, if we need to change an Arduino pin, we just have to make a change at one place, instead of making changes at multiple places in the code. It is a good practice to use variables instead of directly using pin numbers in the code. In the `setup` function, we are setting pins as OUTPUT. If we don't initialize the port to either INPUT or OUTPUT, a port might be in an indefinite state. So, it will give random output.

We have completed the code for one direction of our traffic light. Similarly, you can create your code for the other remaining directions.

LED fading

You can fade out and fade in the light of an LED using Arduino's `analogWrite(pin, value)` function. Before we get into using the `analogWrite()` function, we will understand the concept behind the `analogWrite()` function. To create an analog signal, Arduino uses a technique called **Pulse width modulation (PWM)**.

Pulse width modulation (PWM)

PWM is a technique for getting an analog signal using digital means. By varying the duty cycle (duty cycle is the percentage of a period, when a signal is active.), we can mimic an "average" analog voltage. As you can see in the following image, when we want medium voltage, we will keep the duty cycle as 50%. Similarly, if we want to achieve low voltage and high voltage, we will keep the duty cycle as 10% and 90% respectively. In this application, PWM is the process to control the power sent to the LED:

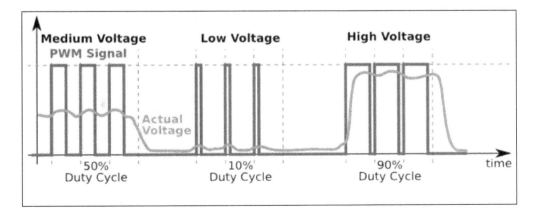

Using PWM on Arduino

Arduino UNO has 14 digital I/O pins. As mentioned in *Chapter 1*, *Getting Started with Arduino and LEDs*, we can use six pins (3, 5, 6, 9, 10, and 11) as PWM pins. These pins are controlled by on-chip timers, which toggle the pins automatically at a rate of about 490 Hz. As discussed earlier, we will be using the `analogWrite()` function.

In the following figure, the `analogWrite()` function takes the pin number and pin value as its parameter. Here, as you can see in the image, the pin value can be between 0 and 255, with the duty cycle mapped to 0% and 100%:

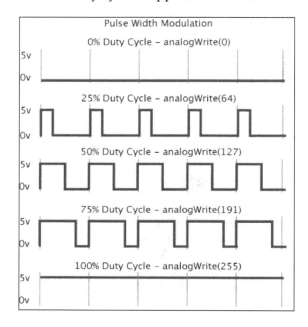

Connect the anode (longer head) of the LED to pin 11 (PWM pin) through a 220 Ω resistor and the cathode (shorter head) to ground, and write the following code in your editor window:

```
int led = 11;            // the pin that the LED is attached to
int brightness = 0;      // how bright the LED is
int steps = 5;      // how many points to fade the LED by

void setup() {
 pinMode(led, OUTPUT);
}
```

```
void loop() {
  // Setting brightness of LED:
  analogWrite(led, brightness);

  // change the brightness for next time through the loop:
  brightness = brightness + steps;

  // When brightness value reaches either 0 or 255, reverse direction
of fading
  if (brightness == 0 || brightness == 255) {
    steps = -steps ;
  }
  // wait for 30 milliseconds to see the dimming effect
  delay(30);
}
```

We are using pin 11(PWM pin) for fading the LED. We are storing the brightness of the LED in variable brightness. Initially, we are setting 0 brightness to the LED. When the `loop` function runs again, we are incrementing the value by steps of 5. As in the `analogWrite()` function, we can set the value between 0 and 255. Once brightness reaches maximum, we are decrementing the value. Similarly, once brightness reaches 0, we start incrementing brightness in steps of 1. To see the dimming effect, we are putting a delay of 30 milliseconds at the end of the code.

Creating a mood lamp

Lighting is one of the biggest opportunities for homeowners to effectively influence the ambience of their home, whether for comfort and convenience or to set the mood for guests. In this section, we will make a simple yet effective mood lamp using our own Arduino. We will be using an RGB LED for creating our mood lamp. An RGB (red, green, and blue) LED has all three LEDs in one single package, so we don't need to use three different LEDs for getting different colors. Also, by mixing the values, we can simulate many colors using some sophisticated programming. It is said that, we can produce 16.7 million different colors.

Using an RGB LED

An RGB LED is simply three LEDs crammed into a single package. An RGB LED has four pins. Out of these four pins, one pin is the cathode (ground). As an RGB LED has all other three pins shorted with each other, it is also called a common anode RGB LED:

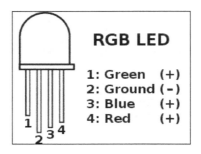

Here, the longer head is the cathode, which is connected with ground, and the other three pins are connected with the power supply. Be sure to use a current-limiting resistor to protect the LED from burning out. Here, we will mix colors as we mix paint on a palette or mix audio with a mixing board. But to get a different color, we will have to write a different analog voltage to the pins of the LED.

Why do RGB LEDs change color?

As your eye has three types of light interceptor (red, green, and blue), you can mix any color you like by varying the quantities of red, green, and blue light. Your eyes and brain process the amounts of red, green, and blue, and convert them into a color of the spectrum:

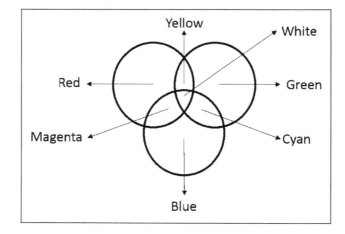

If we set the brightness of all our LEDs the same, the overall color of the light will be white. If we turn off the red LED, then only the green and blue LEDs will be on, which will make a cyan color. We can control the brightness of all three LEDs, making it possible to make any color. Also, the three different LEDs inside a single RGB LED might have different voltage and current levels; you can find out about them in a datasheet. For example, a red LED typically needs 2 V, while green and blue LEDs may drop up to 3-4 V.

Designing a mood lamp

Now, we are all set to use our RGB LED in our mood lamp. We will start by designing the circuit for our mood lamp. In our mood lamp, we will make a smooth transition between multiple colors.

For that, we will need following components:

- An RGB LED
- 270 Ω resistors (for limiting the current supplied to the LED)
- A breadboard

As we did earlier, we need one pin to control one LED. Here, our RGB LED consists of three LEDs. So, we need three different control pins to control three LEDs. Similarly, three current-limiting resistors are required for each LED. Usually, this resistor's value can be between 100 Ω and 1000 Ω. If we use a resistor with a value higher than 1000 Ω, minimal current will flow through the circuit, resulting in negligible light emission from our LED. So, it is advisable to use a resistor having suitable resistance. Usually, a resistor of 220 Ω or 470 Ω is preferred as a current-limiting resistor.

As discussed in the earlier section, we want to control the voltage applied to each pin, so we will have to use PWM pins (3, 5, 6, 9, 10, and 11). The following schematic controls the red LED from pin 11, the blue LED from pin 10, and the green LED from pin 9. Hook the following circuit using resistors, the breadboard, and your RGB LED:

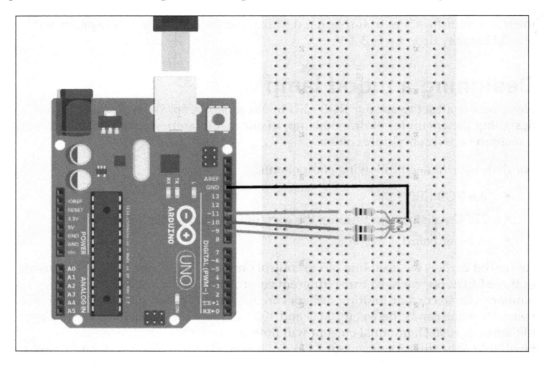

Once you have made the connection, write the following code in the editor window of Arduino IDE:

```
int redLed = 11;
int blueLed = 10;
int greenLed = 9;

void setup()
{
  pinMode(redLed, OUTPUT);
  pinMode(blueLed, OUTPUT);
  pinMode(greenLed, OUTPUT);
}
void loop()
{
  setColor(255, 0, 0); // Red
  delay(500);
```

```
      setColor(255, 0, 255); // Magenta
      delay(500);
      setColor(0, 0, 255); // Blue
      delay(500);
      setColor(0, 255, 255); // Cyan
      delay(500);
      setColor(0, 255, 0); // Green
      delay(500);
      setColor(255, 255, 0); // Yellow
      delay(500);
      setColor(255, 255, 255); // White
      delay(500);
}
void setColor(int red, int green, int blue)
{
      // For common anode LED, we need to subtract value from 255.
      red = 255 - red;
      green = 255 - green;
      blue = 255 - blue;
      analogWrite(redLed, red);
      analogWrite(greenLed, green);
      analogWrite(blueLed, blue);
}
```

We are using very simple code for changing the color of the LED at every one second interval. Here, we are setting the color every second. So, this code won't give you a smooth transition between colors. But with this code, you will be able to run the RGB LED. Now we will modify this code to smoothly transition between colors. For a smooth transition between colors, we will use the following code:

```
int redLed = 11;
int greenLed = 10;
int blueLed = 9;

int redValue   = 0;
int greenValue = 0;
int blueValue  = 0;

void setup(){
  randomSeed(analogRead(0));
}

void loop() {
  redValue = random(0,256); // Randomly generate 1 to 255
  greenValue = random(0,256); // Randomly generate 1 to 255
  blueValue = random(0,256); // Randomly generate 1 to 255
```

```
    analogWrite(redLed,redValue);
    analogWrite(greenLed,greenValue);
    analogWrite(blueLed,blueValue);

// Incrementing all the values one by one after setting the random
values.
    for(redValue = 0; redValue < 255; redValue++){
      analogWrite(redLed,redValue);
      analogWrite(greenLed,greenValue);
      analogWrite(blueLed,blueValue);
      delay(10);
    }
    for(greenValue = 0; greenValue < 255; greenValue++){
      analogWrite(redLed,redValue);
      analogWrite(greenLed,greenValue);
      analogWrite(blueLed,blueValue);
      delay(10);
    }
    for(blueValue = 0; blueValue < 255; blueValue++){
      analogWrite(redLed,redValue);
      analogWrite(greenLed,greenValue);
      analogWrite(blueLed,blueValue);
      delay(10);
    }

    //Decrementing all the values one by one for turning off all the
LEDs.
    for(redValue = 255; redValue > 0; redValue--){
      analogWrite(redLed,redValue);
      analogWrite(greenLed,greenValue);
      analogWrite(blueLed,blueValue);
      delay(10);
    }
    for(greenValue = 255; greenValue > 0; greenValue--){
      analogWrite(redLed,redValue);
      analogWrite(greenLed,greenValue);
      analogWrite(blueLed,blueValue);
      delay(10);
    }
    for(blueValue = 255; blueValue > 0; blueValue--){
      analogWrite(redLed,redValue);
      analogWrite(greenLed,greenValue);
      analogWrite(blueLed,blueValue);
      delay(10);
    }
}
```

We want our mood lamp to repeat the same sequence of colors again and again. So, we are using the `randomSeed()` function. The `randomSeed()` function initializes the pseudo random number generator, which will start at an arbitrary point and will repeat in the same sequence again and again. This sequence is very long and random, but will always be the same. Here, pin 0 is unconnected. So, when we start our sequence using `analogRead(0)`, it will give some random number, which is useful in initializing the random number generator with a pretty fair random number. The `random(min,max)` function generates the random number between min and max values provided as parameters. In the `analogWrite()` function, the number should be between 0 and 255. So, we are setting min and max as 0 and 255 respectively. We are setting the random value to redPulse, greenPulse, and bluePulse, which we are setting to the pins. Once a random number is generated, we increment or decrement the value generated with a step of 1, which will smooth the transition between colors.

Now we are all set to use this as mood lamp in our home. But before that we need to design the outer body of our lamp. We can use white paper (folded in a cube shape) to put around our RGB LED. White paper acts as a diffuser, which will make sure that the light is mixed together. Alternatively, you can use anything which diffuses light and make things looks beautiful! If you want to make the smaller version of the mood lamp, make a hole in a ping pong ball. Extend the RGB LED with jump wires and put that LED in the ball and you are ready to make your home look beautiful.

Developing an LED night lamp

So now we have developed our mood lamp, but it will turn on only when we connect a power supply to Arduino. It won't turn on or off depending on the darkness of the environment. Also, to turn it off, we have to disconnect our power supply from Arduino. In this section, we will learn how to use switches with Arduino.

Introduction to switch

Switches are one of the most elementary and easy-to-overlook components. Switches do only one thing: either they open a circuit or short circuit. Mainly, there are two types of switches:

- **Momentary switches**: Momentary switches are those switches which require continuous actuation—like a keyboard switch and reset button on the Arduino board

- **Maintained switches**: Maintained switches are those switches which, once actuated, remain actuated—like a wall switch.

Normally, all the switches are NO (normally opened) type switches. So, when the switch is actuated, it closes the path and acts as a perfect piece of conducting wire. Apart from this, based on their working, many switches are out there in the world, such as toggle, rotary, DIP, rocker, membrane, and so on.

Here, we will use a normal push button switch with four pins:

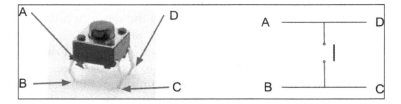

In our push button switch, contacts A-D and B-C are short. We will connect our circuit between A and C. So, whenever you press the switch, the circuit will be complete and current will flow through the circuit. We will read the input from the button using the `digitalRead()` function. We will connect one pin (pin A) to the 5 V, and the other pin (pin C) to Arduino's digital input pin (pin 2). So whenever the key is pressed, it will send a 5 V signal to pin 2.

Pixar lamp

We will add a few more things in the mood lamp we discussed to make it more robust and easy to use. Along with the switch, we will add some kind of light-sensing circuit to make it automatic. We will use a **Light Dependent Resistor (LDR)** for sensing the light and controlling the lamp.

Basically, LDR is a resistor whose resistance changes as the light intensity changes. Mostly, the resistance of LDRs drops as light increases. For getting the value changes as per the light levels, we need to connect our LDR as per the following circuit:

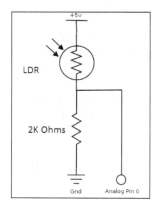

Here, we are using a voltage divider circuit for measuring the light intensity change. As light intensity changes, the resistance of the LDR changes, which in turn changes the voltage across the resistor. We can read the voltage from any analog pin using `analogRead()`.

Once you have connected the circuit as shown, write the following code in the editor:

```
int LDR = 0; //will be getting input from pin A0
int LDRValue = 0;
int light_sensitivity = 400;    //This is the approx value of light
surrounding your LDR
int LED = 13;

void setup()
{
  Serial.begin(9600);                //start the serial monitor with 9600
buad
  pinMode(LED, OUTPUT);

}

void loop()
{
  LDRValue = analogRead(LDR);      //Read the LDR's value through LDR
pin A0
  Serial.println(LDRValue);        //Print the LDR values to serial
monitor

  if (LDRValue < light_sensitivity)
  {
    digitalWrite(LED, HIGH);
  }
  else
  {
    digitalWrite(LED, LOW);
  }
  delay(50);          //Delay before LDR value is read again
}
```

In the preceding code, we are reading the value from our LDR at pin analog A0. Whenever the value read from pin A0 is below a certain threshold value, we are turning on the LED. So whenever the light (lux value) around the LDR drops, then the set value, it will turn on the LED, or in our case, the mood lamp.

Similarly, we will add a switch in our mood lamp to make it fully functional as a pixar lamp.

Connect one pin of the push button at 5 V and the other pin to digital pin 2. We will turn on the lamp only, and only when the room is dark and the switch is on. So we will make the following changes in the previous code.

In the `setup` function, initialize pin 2 as input, and in the loop add the following code:

```
buttonState = digitalRead(pushSwitch); //pushSwitch is initialized as
2.
If (buttonState == HIGH){

//Turn on the lamp
}
Else {
//Turn off the lamp.
//Turn off all LEDs one by one for smoothly turning off the lamp.
  for(redValue = 255; redValue > 0; redValue--){
    analogWrite(redLed,redValue);
    analogWrite(greenLed,greenValue);
    analogWrite(blueLed,blueValue);
    delay(10);
  }
  for(greenValue = 255; greenValue > 0; greenValue--){
    analogWrite(redLed,redValue);
    analogWrite(greenLed,greenValue);
    analogWrite(blueLed,blueValue);
    delay(10);
  }
  for(blueValue = 255; blueValue > 0; blueValue--){
    analogWrite(redLed,redValue);
    analogWrite(greenLed,greenValue);
    analogWrite(blueLed,blueValue);
    delay(10);
  }
}
```

So, now we have incorporated an LDR and switch in our lamp to use it as a normal lamp.

Summary

In this chapter, we started with using a breadboard. Programming multiple LEDs was explained by developing a model traffic light controller. A mood lamp with RGB LED and a Pixar lamp with a sensing part were developed by the end of this chapter.

In the next chapter, a remote-controlled TV backlight will be developed, where you will learn how to make communication between Arduino and a TV remote.

3

Project 2 – Remote Controlled TV Backlight

In the previous chapter, *Project 1 – LED Night Lamp*, we dived into the amazing world of LEDs. We created some cool projects using different types of LEDs. Now, we will use another type of LED – an IR (Infrared) LED. In this chapter, we will start by learning the basics of IR LEDs and the basics of IR communication. Once you have learned about programming the IR sensor, we will use it to control a TV backlight using a remote. In this chapter, you will learn about the following:

- Introduction to and the workings of an IR LED
- Programming an IR sensor
- How to control an LED array
- Developing a remote controlled TV backlight

Introduction to IR LEDs

In the world of wireless technology, IR (infrared) is one of the most common, inexpensive, and easy to use modes of communication. You might have always wondered how a TV remote works. A TV remote uses IR LEDs to send out the signal. As the wavelength of light emitted from the IR LED is longer than the visible light, one can't see it with the naked eye. But, if you look through the camera of your mobile or any other camera, you can see the light beaming when you press any key on the remote. Let's first understand what an IR LED is and what the different applications of an IR LED are.

What is IR LED?

An IR (infrared) LED, also known as an IR (infrared) transmitter, transmits infrared waves in the range of 875 nm to 950 nm. Usually, IR LEDs are made up of gallium arsenide or aluminum gallium arsenide. The working principle of an IR LED is the same as we mentioned in the previous chapters. The longer lead of the LED is the anode and the shorter one is the cathode, as shown here:

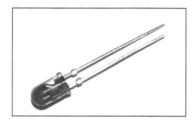

Applications of IR LED / IR communication

Apart from using IR communication in TV remote controls, there are a number of other applications that use IR communication. Infrared light can also be used to transfer data between electronic devices. Although infrared light is invisible to the eye, digital cameras and other sensors can see this light, so infrared light can be used for many security systems and night vision technology. Apart from technical uses, the U.S. Food and Drug Administration Department has approved several products with IR LEDs for use in some medical and cosmetic procedures.

IR sensors

We have learned the basics of IR communication and IR LEDs, so now we will move on to making our own IR sensor.

There are various types of IR sensors, depending upon the application. Proximity sensors, obstruction sensors, and contrast sensors (used in line follower robot) are a few examples which use IR sensors.

Working mechanism

An IR sensor consists of an IR LED (which works as a transmitter) and a photodiode (which works as a receiver) pair. The IR LED emits IR radiation, which, on receiving at the photodiode dictates the output of the sensor.

There are different types of use cases for an IR sensor. For example, if we held an IR LED directly in front of the photodiode, such that almost all the radiation reaches the photodiode, this can be used as burglar alarm. So, when anyone interrupts the line of sight between the IR LED and the photodiode, this will break the continuous radiation coming from the IR LED, and we can program it to raise an alarm. This type of mechanism is also called a direct incidence, as we are not using any reflective surface in between the transmitter and receiver.

Another use case is with indirect incidence, in which we use physics' law of absorption. This means, when light is directed at a black surface, the black surface will actually absorb the light. Similarly, when IR radiation is directed at a black surface, the surface will absorb the IR radiation. But, when it is directed toward a white surface, the surface will reflect the IR radiation. Based on the amount of light received back from the surface, we can detect, if robot is following a line or not. So, the absorption principle can be used for line follower robot:

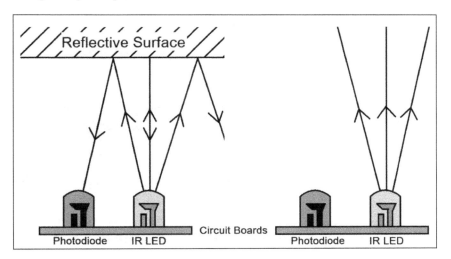

As you can see in the preceding image, whenever any obstacle is detected in the path of the IR, some of the IR radiation is reflected back, which, on receipt of the IR waves, gives the output. This type of setup is also useful for detecting any object/obstacle in the path.

Programming a basic IR sensor

After understanding the basic workings of a simple IR sensor, we will learn how to program this sensor.

Based on the principle of direct incidence, we will develop a simple burglar alarm.

Once you have connected the circuit as shown in the following diagram, upload the following code to Arduino:

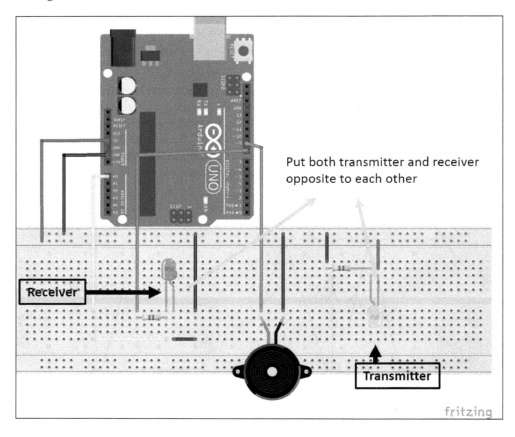

```
int IRTransmitter = 8;
int IRReceiver = A0; //IR Receiver is connected to pin 5

int buzzer = 9; // Buzzer is connected with pin 9.
int output = 0; // Variable to store the value from the IR sensor
int ambientLight = 500;

void setup()
{
  pinMode(IRReceiver,INPUT);
  pinMode(IRTransmitter,OUTPUT);
  pinMode(buzzer,OUTPUT);
  digitalWrite(IRTransmitter,HIGH);
}
```

```
void loop()
{
  output = analogRead(IRReceiver);
  // If anything comes in between the IR Transmitter and IR receiver
  // IR receiver will not give the output. Make an alarm.
  if (output < ambientLight)
  {
    makeAlarm(50);
  }
}
void makeAlarm(unsigned char time)
{
  analogWrite(buzzer,170);
  delay(time);
  analogWrite(buzzer,0);
  delay(time);
}
```

In the preceding code, we are continuously making the IR transmitter ON. So, the IR receiver/photodiode will continuously receive an IR signal. Whenever any person tries to break into the house or safe, the IR signal will get interrupted, which in turn will lower the voltage drop across the IR receiver and an alarm will be raised. Here, based on the conditions around you, you will have to change the value of the ambientLight variable in the code.

How to receive data from a TV remote

As we discussed earlier, a TV remote has an IR LED, which transmits IR light with a 38 kHz frequency. So, whenever that radiation is detected at the receiver part of the TV (the control circuit of the TV), we need to filter that signal.

As you can see in the following image, when a signal is actually transmitted from the TV remote, it will look like a series of waves:

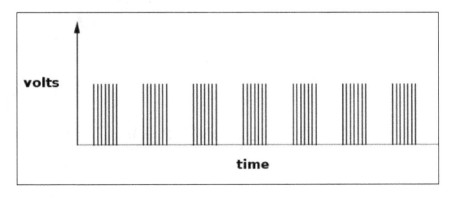

For filtering the actual signal from a TV remote from ambient noise, we will use TSOP38238 IC. TSOP38238 looks like a transistor, but actually this device combines an IR sensitive photocell, a 38 kHz band pass filter, and an automatic gain controller. Here, the IR sensitive photocell works as an IR receiver (photodiode), and the 38 kHz band pass filter is required to smooth the received modulated signal.

After passing through the 38 kHz band pass filter, the output signal will look as shown in the following, which is much cleaner and easier to read:

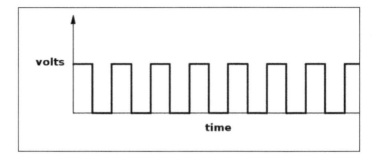

Also, TSOP38238 is covered in the lead package epoxy frame, which acts as an IR filter. So, the disturbance light (DC light from a tungsten bulb, or a modulated signal from fluorescent lamps) does not reach the photodiode. This demodulated signal can be directly decoded by any microprocessor (in our case, Arduino). Because of this, apart from preamplifier, it ignores all other IR light unless it is modulated at a specific frequency (38 kHz).

This IC is as simple as an IC can get with its three pins. There are two power pins and one pin for the output signal:

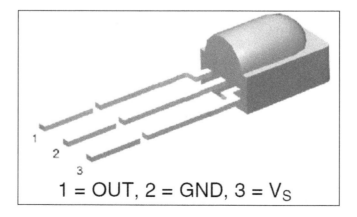

TSOP38238 can be powered with a supply from 2.5 V to 5.5 V, which makes it suitable for all types of application. Also, it can receive signals from almost all types of remotes.

To start with, we will control a single LED using an IR remote. For that, we will be using the IRRemote library, available at https://github.com/z3t0/Arduino-IRremote. Once you have downloaded and extracted the code in the libraries folder of Arduino, you will be able to see the examples in the Arduino IDE. We need to record the value of all the buttons for future use.

Connect the first pin of TSOP38238 to pin 11 of Arduino, the second pin to the ground pin of Arduino, and the third pin to the 5 V pin of Arduino. Connect Arduino over USB and try to compile the following code:

```
#include <IRremote.h>
const int IR_RECEIVER = 11;
IRrecv receiver(IR_RECEIVER);
decode_results buttonPressed;
void setup()
{
  Serial.begin(9600);
  receiver.enableIRIn(); // Start the receiver
}

void loop()
{
  if (receiver.decode(&buttonPressed))
  {
    Serial.println(buttonPressed.value); //Printing values coming from
the IR Remote
    receiver.resume(); // Receive the next value
  }
  delay(100);
}
```

 In the latest version of Arduino, the IRremote.h library is already installed in the RobotIRremote folder. But, you won't have all the examples available from the downloaded library. So, delete the RobotIRremote library and try to compile the code again.

Once you have deleted the duplicate file, you should be able to successfully compile the code. After uploading the preceding code, open the serial monitor. As we want to control a single LED using a remote, we will have to know the value of the button pressed. So, try noting down the values of all the buttons one after the other.

For example, digit 1 has 54,528 codes for a certain remote. You might have to check for the remote that you have. We will now control the LED by using the IR remote. Along with the IR receiver, we will connect one LED, as shown in the following circuit:

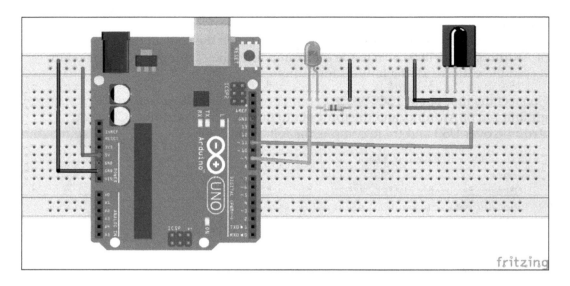

Update the code for controlling the LED, based on your readings from the previous exercise:

```
#include <IRremote.h>
const int IR_RECEIVER = 11;
IRrecv receiver(IR_RECEIVER);
decode_results buttonPressed;
long int buttonValue = 0;
const long int buttonOne = 54528; //Update value one according to your
readings
const long int buttonTwo = 54572; //Update value two according to your
readings


int LEDpin = 9;

void setup()
{
  Serial.begin(9600);
  receiver.enableIRIn(); // Start the receiver
  pinMode(LEDpin,OUTPUT);
}

void loop()
{
```

```
    if (receiver.decode(&buttonPressed))
    {
      buttonValue = buttonPressed.value;
      Serial.println(buttonValue);
      switch (buttonValue){
        case buttonOne:
            digitalWrite(LEDpin,HIGH);
            break;
        case buttonTwo:
            digitalWrite(LEDpin,LOW);
            break;
        default:
            Serial.println("Waiting for input. ");
      }
      receiver.resume(); // Receive the next value
    }
    delay(100);
}
```

Now you will be able to control a single LED using the IR remote. In a similar way, you can control anything that you want to control using an IR remote. At the end of this chapter, we want to control a TV backlight using an IR remote. We will develop a TV backlight using an LED strip. So, we will now learn about how to control an LED array.

LED strips

LED strips are flexible circuit boards with full color LEDs soldered on them. There are two basic types of LED strip: the "analog" and "digital" kinds. Analog strips have all the LEDs connected in parallel. So, they act as one huge tri-color LED. In case of an analog LED strip, we can't set the color/brightness of each LED. So, they are easy to use and inexpensive. Digital LED strips work in a different way. To use the LED, we have to send digital code corresponding to each LED in the case of a digital LED strip. As they provide more modularity, they are quite expensive. Also, digital LED strips are difficult to use compared to analog LED strips:

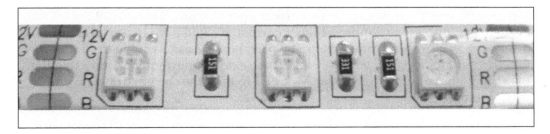

Internally, RGB LEDs are connected to each other in parallel. One section of the strip contains all three LEDs, connected in parallel. A complete LED strip is made up of a number of parallel RGB LEDs connected in series. Connection in one block/section is as shown in the following figure:

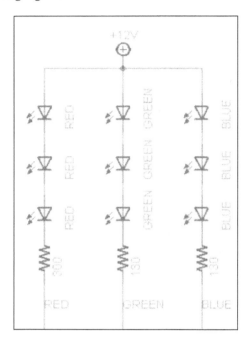

As one section contains multiple LEDs, it requires more current to run. It can't run on the current provided from Arduino. In case of full white, each segment will require approximately 60 mA of current per block/section. So to run the complete LED strip, we may require current of up to 1.2 A per meter. For running an LED strip, an external power supply of 12 V is required.

Controlling an LED strip with Arduino

As we know, we can draw maximum current up to 300 mA if we put all the I/O pins together. But, as we require a higher current, we will have to use an external power supply. Now here comes the tricky part. We will use Arduino to control the LED strip, and for a power supply, we will use external power. For connecting the LED strip with Arduino, we will use MOSFET to limit the current drawn to and from Arduino.

Here, we will use N-channel MOSFET, which is very inexpensive and works with 3.3 V to 5 V logic. These FETs can switch over 60 A and 30 V.

Connecting to the strip is relatively easy. We need to solder the wires to the four pins/copper pads of our LED strip. One can use heat shrink for providing insulation, abrasion resistance, and environmental protection.

To use our LED strip with Arduino, we will require some resistors as well, to limit the current:

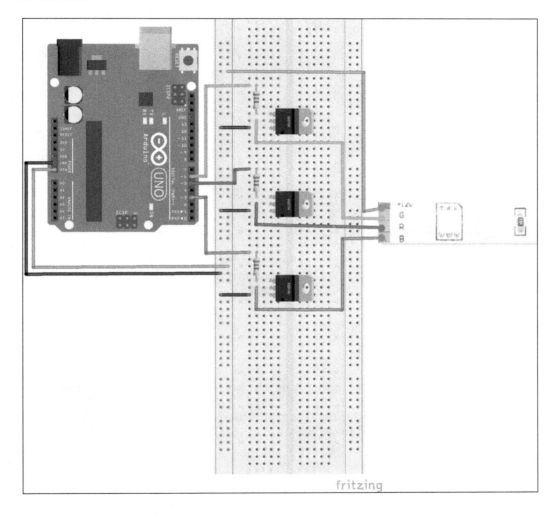

Connect the power transistor with the resistor in between the PWM output pin and the base. In case of NPN transistors, pin 1 is the base. Pin 2 is the collector and pin 3 is the emitter. Now, for power supply to Arduino and the LED strip, connect a 9 -12 V power supply to the Arduino, so that Vin supplies the high voltage to the LED.

At the end of the wiring connection, make sure to connect the ground of the supply to that of Arduino/MOSFETs. Connect the Arduino output pin to the base of the MOSFET (pin 1) with a 100-220 Ω resistor in between. Connect pin 2 of the MOSFET to the LED strip's input pin and connect pin 3 of the MOSFET to the ground.

Check all the connections and write the following code in the editor window of Arduino. Here, we are using pins 5, 6, and 3 of Arduino to control the LED strip's red, green, and blue LEDs respectively:

```
int redLed = 5;
int greenLed = 6;
int blueLed = 3;

int redValue   = 0;
int greenValue = 0;
int blueValue  = 0;

void setup(){
  randomSeed(analogRead(0));
}

void loop() {
  redValue = random(0,256); // Randomly generate 1 to 255
  greenValue = random(0,256); // Randomly generate 1 to 255
  blueValue = random(0,256); // Randomly generate 1 to 255

  analogWrite(redLed,redValue);
  analogWrite(greenLed,greenValue);
  analogWrite(blueLed,blueValue);

// Incrementing all the values one by one after setting the random
values.
  for(redValue = 0; redValue < 255; redValue++){
    analogWrite(redLed,redValue);
    analogWrite(greenLed,greenValue);
    analogWrite(blueLed,blueValue);
    delay(10);
  }
  for(greenValue = 0; greenValue < 255; greenValue++){
    analogWrite(redLed,redValue);
    analogWrite(greenLed,greenValue);
    analogWrite(blueLed,blueValue);
```

```
      delay(10);
    }
    for(blueValue = 0; blueValue < 255; blueValue++){
      analogWrite(redLed,redValue);
      analogWrite(greenLed,greenValue);
      analogWrite(blueLed,blueValue);
      delay(10);
    }

    //Decrementing all the values one by one for turning off all the
LEDs.
    for(redValue = 255; redValue > 0; redValue--){
      analogWrite(redLed,redValue);
      analogWrite(greenLed,greenValue);
      analogWrite(blueLed,blueValue);
      delay(10);
    }
    for(greenValue = 255; greenValue > 0; greenValue--){
      analogWrite(redLed,redValue);
      analogWrite(greenLed,greenValue);
      analogWrite(blueLed,blueValue);
      delay(10);
    }
    for(blueValue = 255; blueValue > 0; blueValue--){
      analogWrite(redLed,redValue);
      analogWrite(greenLed,greenValue);
      analogWrite(blueLed,blueValue);
      delay(10);
    }
  }
}
```

If everything is in place, you should be able to see the LED strip getting on and changing its color.

Now we have learned about all the things required to control the TV backlight using IR remote, so we will integrate all the things that we learned in this chapter:

- How IR works
- How to read values from an IR remote
- How to control an LED strip

We want to control the brightness of the LED strip as well. We will use the power button to turn the backlight off and on. With volume plus and volume minus, we will increase and decrease the brightness of the backlight.

As we did earlier in the chapter, connect TSOP38238 to pin 11, 5 V, and ground pin of Arduino. Once you have done all the connections, upload the following code on Arduino:

```
#include <IRremote.h>
const int IR_RECEIVER = 11; // Connect output pin of TSOP38238 to pin
11
IRrecv receiver(IR_RECEIVER);
decode_results buttonPressed;
long int buttonValue = 0;

// Mention the codes, you get from previous exercise
const long int POWER_BUTTON = 54572; // Power button to turn on or off
the backlight
const long int PLUS_BUTTON = 54536;  // Increase brightness of the LED
Strip
const long int MINUS_BUTTON = 54608; // Decrease brightness of the LED
strip
const long int CHANGE_COLOR = 54584; // Decrease brightness of the LED
strip

const int FADE_AMOUNT = 5; // For fast increasing/decreasing
brightness increase this value
boolean isOn = false;

int redLed = 5;
int greenLed = 6;
int blueLed = 3;

int redValue   = 0;
int greenValue = 0;
int blueValue  = 0;

int colors[3];

// Power up the LED strip with random color
void powerUp(int *colors)
{
  redValue = random(0, 256); // Randomly generate 1 to 255
  greenValue = random(0, 256); // Randomly generate 1 to 255
  blueValue = random(0, 256); // Randomly generate 1 to 255

  analogWrite(redLed, redValue);
  analogWrite(greenLed, greenValue);
```

```
  analogWrite(blueLed, blueValue);

  colors[0] = redValue;
  colors[1] = greenValue;
  colors[2] = blueValue;
}

// Turn off the LED
void powerDown(int *colors)
{
  redValue = colors[0];
  greenValue = colors[1];
  blueValue = colors[2];

  //Decrementing all the values one by one for turning off all the
LEDs.
  for (; redValue > 0; redValue--) {
    analogWrite(redLed, redValue);
    delay(10);
  }
  for (; greenValue > 0; greenValue--) {
    analogWrite(greenLed, greenValue);
    delay(10);
  }
  for (; blueValue > 0; blueValue--) {
    analogWrite(blueLed, blueValue);
    delay(10);
  }
  colors[0] = redValue;
  colors[1] = greenValue;
  colors[2] = blueValue;
}

void increaseBrightness(int *colors)
{
  redValue = colors[0];
  greenValue = colors[1];
  blueValue = colors[2];

  redValue += FADE_AMOUNT;
  greenValue += FADE_AMOUNT;
  blueValue += FADE_AMOUNT;

  if (redValue >= 255) {
```

```
    redValue = 255;
  }

  if (greenValue >= 255) {
    greenValue = 255;
  }

  if (blueValue >= 255) {
    blueValue = 255;
  }
  analogWrite(redLed, redValue);
  analogWrite(greenLed, greenValue);
  analogWrite(blueLed, blueValue);

  colors[0] = redValue;
  colors[1] = greenValue;
  colors[2] = blueValue;
}

void decreaseBrightness(int *colors)
{
  redValue = colors[0];
  greenValue = colors[1];

  blueValue = colors[2];

  redValue -= FADE_AMOUNT;
  greenValue -= FADE_AMOUNT;
  blueValue -= FADE_AMOUNT;

  if (redValue <= 5) {
    redValue = 0;
  }

  if (greenValue <= 5) {
    greenValue = 0;
  }

  if (blueValue <= 5) {
    blueValue = 0;
  }
  analogWrite(redLed, redValue);
  analogWrite(greenLed, greenValue);
```

```
  analogWrite(blueLed, blueValue);

  colors[0] = redValue;
  colors[1] = greenValue;
  colors[2] = blueValue;
}

// Randomly generates a color and make a smooth transition to that
color
void changeColor(int *colors)
{
  int newRedValue = random(0, 256); // Randomly generate 1 to 255
  int newGreenValue = random(0, 256); // Randomly generate 1 to 255
  int newBlueValue = random(0, 256); // Randomly generate 1 to 255

  redValue = colors[0];
  greenValue = colors[1];
  blueValue = colors[2];

  if (newRedValue > redValue) {
    for (; redValue >= newRedValue; redValue++) {
      analogWrite(redLed, redValue);
      delay(10);
    }
  }
  else {
    for (; redValue <= newRedValue; redValue--) {
      analogWrite(redLed, redValue);
      delay(10);
    }
  }

  if (newGreenValue > greenValue) {
    for (; greenValue >= newGreenValue; greenValue++) {
      analogWrite(greenLed, greenValue);
      delay(10);
    }
  }
  else {
    for (; greenValue <= newGreenValue; greenValue--) {
      analogWrite(greenLed, greenValue);
      delay(10);
    }
  }
```

```
  if (newBlueValue > blueValue) {
    for (; blueValue >= newBlueValue; blueValue++) {
      analogWrite(blueLed, blueValue);
      delay(10);
    }
  }
  else {
    for (; blueValue <= newBlueValue; blueValue--) {
      analogWrite(blueLed, blueValue);
      delay(10);
    }
  }

  colors[0] = redValue;
  colors[1] = greenValue;
  colors[2] = blueValue;
}

void setup() {
  Serial.begin(9600);
  receiver.enableIRIn(); // Start the receiver

  randomSeed(analogRead(0));

  pinMode(redLed, OUTPUT);
  pinMode(greenLed, OUTPUT);
  pinMode(blueLed, OUTPUT);
}

void loop() {

  if (receiver.decode(&buttonPressed))
  {
    buttonValue = buttonPressed.value;
    Serial.println(buttonValue);
    switch (buttonValue) {
      case POWER_BUTTON:
        if (!isOn) {
          powerUp(colors);
          isOn = true;
        }
        else {
          powerDown(colors);
```

```
        isOn = false;
      }
      break;
    case PLUS_BUTTON:
      decreaseBrightness(colors);
      break;
    case MINUS_BUTTON:
      increaseBrightness(colors);
      break;
    case CHANGE_COLOR:
      changeColor(colors);
      break;
    default:
      Serial.println("Waiting for input. ");
    }
    receiver.resume(); // Receive the next value
  }
  delay(100);
}
```

In the preceding code, we are using the power button to turn the backlight on and off, and the volume up and down buttons to increase and decrease the brightness respectively. I have used the up arrow button to change the color of the strip. You can add more features to this project by configuring it in the switch case block.

Summary

In this chapter, we started with the basics of IR LEDs and IR communication. After that, we learnt about programming IR sensors and their applications. By the end of this chapter, we learnt about controlling an LED strip, and we completed our chapter by developing a remote controlled TV backlight.

In the coming chapters, we will start into the more advanced stuff of developing an LED cube, sound visualization, and persistence of vision.

4
Project 3 – LED Cube

If you have successfully implemented the last two projects, you will have noticed that there is very little or no soldering involved. However, I would say you haven't worked on electronics if you haven't done some intense soldering and burnt your hands. In this chapter, you will get introduced to soldering in detail. You will also understand how to create a 4*4*4 LED cube using an Arduino UNO board. You will learn about the following:

- Introduction to soldering
- Designing an LED cube
- Programming a 4*4*4 LED cube

Getting started with soldering

Soldering is the process of making a sound electrical and mechanical joint between certain metals by joining them with a soft solder. This is an alloy of lead and tin with a low melting point. The joint is heated to the correct temperature by a soldering iron. Effective soldering requires good heat transfer from the iron to the components to be soldered. The longer heat is applied, the greater the risk of heat damage to the wire or component, so it's important to get the job wrapped up quickly.

What you will need

Before you proceed further with the next section of the chapter, make sure you have the following items with you and have got yourself familiarized with the tools:

- Soldering iron
- Basic stand
- Solder desoldering pump
- Cardboard

The following image is for your quick reference so that you will get a rough idea of soldering tools. One thing to remember is that soldering tools vary from place to place, so don't worry if your tools don't look exactly the same. An important thing to note at the beginner level is that your soldering skills are highly dependent on the kind of soldering tools you are using, so make sure you buy the best soldering tools available in your country:

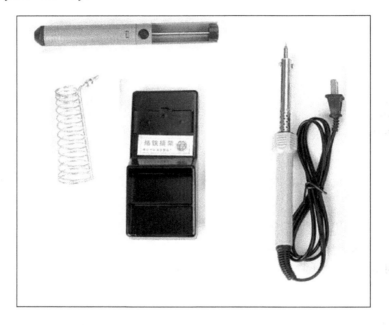

 Go to any electronics components store and ask for a soldering kit. They will give you all the necessary components needed for soldering. Alternatively, you can order electronics components from a few online stores as well.

Safety tips

Soldering poses a few different dangers, so, to stay as safe as possible, always follow these soldering safety tips:

- Never touch the element or tip of the soldering iron. They are very hot (about 400 degree Celsius) and will give you a nasty burn.
- Take great care to avoid touching the mains flex with the tip of the iron.

- Always return the soldering iron to its stand when not in use.
- Never put the soldering iron down on your workbench, even for a moment!
- Work in a well-ventilated area.
- The smoke formed as you melt solder is mostly from the flux and quite irritating. Avoid breathing it in by keeping your head to the side of, not above, your work.
- Wash your hands after using solder.
- Solder contains lead which is a poisonous metal.
- Never solder a live circuit (one that is energized).

Designing an LED cube

As mentioned before, the main focus of this chapter is on soldering. In this section, you will learn how to design an LED cube, which will have intense soldering, and creative elements like LED control and Arduino programming.

Required components

Before getting into the cube design, make sure you have following components for this project:

- Arduino UNO
- 64 LEDs: You can use any color LED. Although 64 LEDs are required for this project, I would recommend you to buy at least 100 LEDs in case some LEDs get burned during the soldering process.
- 16 resistors: These must be appropriate to your LEDs. If you are not sure which resistor to purchase, get 500 Ω/1k Ω resistors.
- Connecting wires
- A printed circuit board
- Thermocol
- Soldering iron and solder wire

Principle behind the design

Before you read this section, make sure you have the listed components or have already ordered them. This section is the most important part of this chapter as it explains the key principle behind the project. It gives a complete overview of the system so that it always stays in the back of your mind and can help identify stuff as you go along. On the Internet, you will find lots of tutorials on making an LED cube, however, most of them would use a single output pin for every single LED. If you use that approach for a 4*4*4 LED cube, you would need 64 pins, which Arduino UNO doesn't have. One approach would be to use shift registers, which will make it complicated for beginners and first-timers.

If you look at the Arduino UNO board, you will notice that at max, 20 pins can be used. So to control all those LEDs in 20 pins, a multiplexing technique will be used. If you break down a cube into four layers, you will need 16 control pins for referencing individual LEDs. To enable a particular LED of the cube, all the layers needs to be activated. So in total, you need 16+4 (20) pins for this project.

There will be a common cathode (negative terminal) for each layer, so all the negative legs of the LEDs are connected to a single pin for that layer. For the anode (positive terminal), each LED will be connected to the corresponding layer above and below. So in total, you will have 16 columns of positive terminals and four layers of negative terminals. Following are some 3D views of the design, which will help you better understand it. Red ones are positive terminals and blue ones are negative terminals:

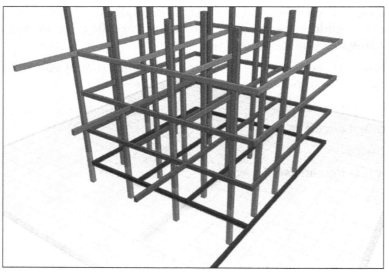

Source: http://cdn.makeuseof.com/wp-content/uploads/2012/07/structural-diagram1.png

Here is another image for your reference:

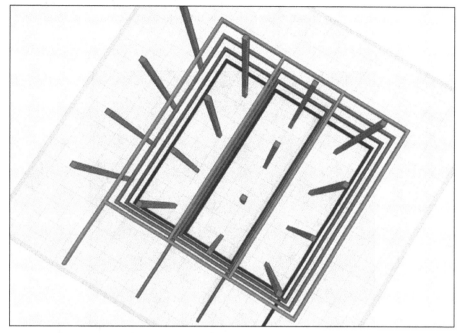

Source: http://cdn.makeuseof.com/wp-content/uploads/2012/07/cube-wiring-layers-from-top.png

Construction

For the construction, you can either use a full metal structure to give rigidity to the structure at the cost of simplicity, or the other, simpler option is to overlap all the legs of the LEDs by about a quarter, which in turn will give the rigidity to the structure that we require. As shown in the following image, fold the cathode of your LEDs:

You can choose to bend the cathode to the left or right. Just make sure that you are consistent across the whole structure and it never touches the anode.

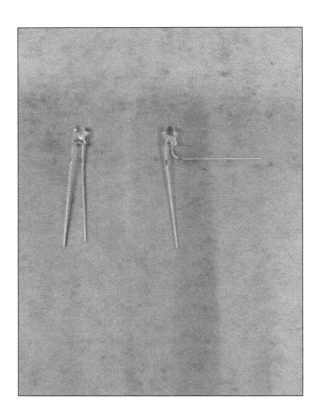

One of the critical parts of this project is making a structure to hold the LED while soldering. You can make this kind of structure by making a wooden jig or using a thermocol sheet. The thermocol sheet option is going to be a bit easier. Here are a couple of things that you should keep in mind while working on this part:

- As mentioned before, it will hold an LED layer so make sure it is accurate and not too loose.
- A quarter of the LED leg should overlap with the adjacent LED. If needed, use a ruler.
- If you are using a wooden jig, make sure your drill is the same size as the LED. If it is a bit tight, you won't be able to remove the fully soldered layer.

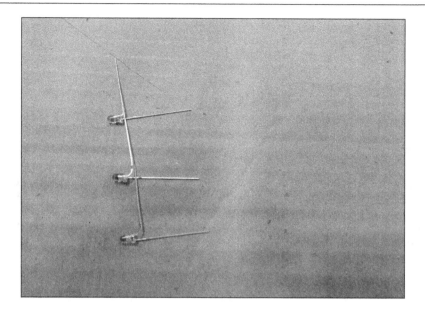

Now solder the cathodes of four rows of LEDs. After completing the first four rows, your structure will look something similar to the following image:

In the preceding image, only two rows of layers are being displayed. If you are working with copper LED cables, you would know that structures created using copper cables won't be that strong. To strengthen the rigidity of the layer, cut and solder two straight bits of craft wire to either end. Make sure they connect with each row. That's it. Your first layer is ready.

Before soldering other layers, you should test the layer that you have just created and correct any errors if you have made any, so that you don't repeat the same mistake while working on other layers. For that, open Arduino IDE and load the blink application that you developed in *Chapter 1, Getting Started with Arduino LEDs*. For testing each LED, connect the ground pin to the layer frame with the connected resistor and press the positive lead to each LED.

If everything goes well, each LED will light up. If not, check the connection for that LED and make sure you haven't missed a solder joint somewhere. If the soldering and connection are right, replace the LED and test it again. There is a chance that the LEDs might have heated up if you kept the soldering iron connected to them for long.

Once you have tested the first layer, remove that layer from the wooden/thermocol jig and repeat the process three more times. Don't forget to test all the four layers of LEDs with the blink program.

Once you have completed and tested the four layers, you need to join all the vertical legs together. One of the techniques is to cut the piece of the cardboard.

It will help to keep all the layers at the same height. You will notice that even after using this there are lots of legs which don't align perfectly. You can use some crocodile pins to hold them in place. Another option is to get some copper cables and solder them at those places where the legs are not aligned properly.

Mistakes to avoid

When you are working with electronics, things can become quite tricky sometimes. Here are a couple of things that you should keep in mind, as most first-timers make mistakes with this:

- If you are using cardboard for getting equal height for the all the layers, make the cardboard longer on the side, and join the pieces of card outside of the cube, so when you've completed the layer, you can easily pull out the card.

- A general rule of thumb is the anode of an LED will connect with the anode of another LED, and similarly, the cathode of an LED will connect with the cathode of another LED. Don't connect/solder the anode of one LED with the cathode of another LED.

Make sure you test all four layers once you have connected them. The easiest way to test is to touch only on the uppermost layer. If that works, it means you have a good connection going through all the layers.

Don't forget to cut extra bits of metal frame and legs before you connect them to the circuit board or cardboard. An important thing to consider here is to not cut vertical legs as those need to put into the prototyping board.

Fixing to the board

Before you move into the Arduino part, the last step is to fix this structure on a prototyping board. If you want to go ahead with a prototyping board, make sure you place each LED at equal distance, and for holding each LED leg you can use some crocodile pins.

Personally, I found it a bit difficult, as once you have joined all the four layers, all the LED legs won't be at an equal distance, so instead I have used a cardboard sheet and soldered the LEDs on cardboard.

The next step is to connect the LEDs with the resistors. Ideally, you should have the resistors soldered on a prototyping board.

If you are using cardboard, as I did, you can use thermocol for connecting a resistor to an LED.

Once you have connected the resistors with the cube, it will look like the following image. Please note, I have used thermocol and cardboard for this prototype. If you are using a wooden base and a prototyping board, your prototype will look much more polished:

If you didn't plan your resistor placement in advance, here is what you will have once you have completed all the steps. A better way is to space them equally in a stepping fashion, so that then you could use one entire side of the cube for all the final connections to Arduino. Here's the circuit diagram:

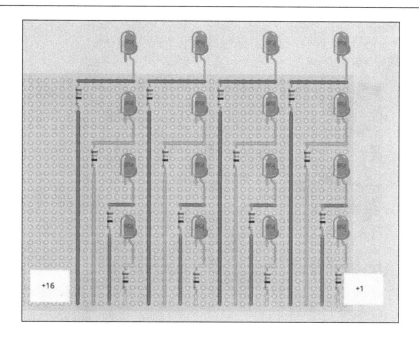

Finally, connect some connecting wires, which can be plugged into relevant Arduino pins, and make sure you use long wires for this. You can use a color code for differentiating between connecting cables:

If you are having trouble getting different color cables and at the end you feel like your soldering and connections are going to be a bit messy, cover your cube internally with a cardboard box, as shown in the following figure:

Programming a 4*4*4 LED cube

Having done the hard part of soldering, let's get into Arduino connection and programming. Before connecting the positive leads, connect four negative layers to Arduino analog I/O ports A2 (bottom layer) through A5 (top layer). After that, 16 LED control pins needs to be connected to the Arduino board. Connect the first 14 pins to Arduino digital I/O ports 0 to 13. The remaining pins 15 and 16 need to be connected to Analog pins A0 and A1. See the following diagram for connection reference:

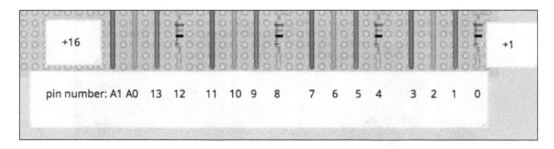

There are a few things that you should understand before programming your cube:

- To address a single LED use a plane (layer) number 0–3, and an LED pin number 0–15. Turn the plane to LOW output (negative leg) and the LED pin number to HIGH (positive leg) to activate the LED.

- Before activating a single LED, ensure all other planes are off by setting them to HIGH output. If you don't do this, the whole column of LEDs will light up instead of a single LED.

Now you are all set to start creating your own programming sequence.

Copy the following code (inspired from `http://www.tecnosalva.com/files/ficheros/ledcube2.ino`) and create a new Arduino sketch and upload the code to your Arduino board:

```
#include <avr/pgmspace.h> // allows use of PROGMEM to store patterns
in flash

#define CUBESIZE 4
#define PLANESIZE CUBESIZE*CUBESIZE
#define PLANETIME 3333 // time each plane is displayed in us -> 100 Hz
refresh
#define TIMECONST 5 // multiplies DisplayTime to get ms

/*
** Defining pins in array makes it easier to rearrange how cube is
wired
** Adjust numbers here until LEDs flash in order - L to R, T to B
** Note that analog inputs 0-5 are also digital outputs 14-19!
** Pin DigitalOut0 (serial RX) and AnalogIn5 are left open for future
apps
*/

int LEDPin[] = {0, 1, 2, 3, 4, 5, 6, 7, 8, 9, 10, 11, 12, 13, 14, 15};
int LEDPinCount = 16;
int PlanePin[] = {16, 17, 18, 19};
int PlanePinCount  = 4;

// initialization
void setup()
{
  int pin; // loop counter
  // set up LED pins as output (active HIGH)
  for (pin=0; pin<PLANESIZE; pin++) {
    pinMode( LEDPin[pin], OUTPUT );
```

```
    }
    // set up plane pins as outputs (active LOW)
    for (pin=0; pin<CUBESIZE; pin++) {
      pinMode( PlanePin[pin], OUTPUT );
    }
}

void loop(){
  loopFor();
}

// the principles of using 4 planes and 16 pins - here we loop over
each, turning on and off in turn
void loopFor()
{
    for(int thisPlane = 0; thisPlane < PlanePinCount; thisPlane++){
      for(int thisPin = 0; thisPin < LEDPinCount; thisPin++){

        planesOff();
        digitalWrite(LEDPin[thisPin],HIGH);
        digitalWrite(PlanePin[thisPlane],LOW);


        delay(50);

        digitalWrite(LEDPin[thisPin],LOW);
        digitalWrite(PlanePin[thisPlane],HIGH);


      }
    }
}

void planesOff(){
    for(int thisPlane = 0; thisPlane < PlanePinCount; thisPlane++){
        digitalWrite(PlanePin[thisPlane],HIGH);
    }
}
```

You would notice that the preceding code simply lights every LED one by one, in sequence. We use two for loops for this, iterating over each layer and each control pin.

That's it! Of course, this is not the only way to control a 4*4*4 LED cube. There are multiple ways you can control LEDs.

Summary

So far in this book, the focus was more on understanding Arduino programming and less about electronics and soldering. If you reached the end of this chapter, this means you have learnt the very important skill that is soldering. Having understood soldering and Arduino programming, you can now bring your own ideas into reality/prototype. Share your prototype and experience on social media using hashtag #ArduinoBLINK.

Having gained sound knowledge of Arduino programming and soldering, in the next chapter you will learn about sound visualization and how to use different sensors with Arduino.

Sound Visualization and LED Christmas Tree

5

Things get pretty easy once you have understood the basics of Arduino and how to control the "stuff" with Arduino. In the previous chapters, we have developed some useful projects using LEDs and light sensors. We also learned soldering in the previous chapter. In this chapter, we will understand how to visualize sound using Arduino and then we will develop an LED Christmas tree.

- Introduction to sound visualization
- Sound visualization using Arduino
- Developing a sound controlled Christmas tree

Introduction to sound visualization

Sound visualization, or music visualization, has been an integral part of the music industry since the evolution of media players. For example, on your computer, when you play any music, you can see the visualization in the default media player or VLC media player, as shown in the following image:

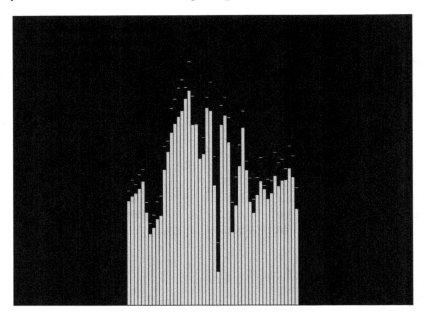

You have noticed that, as the loudness of the music, or the frequency of the sound changes, the visualization changes. We will use the same principle to visualize sound using Arduino.

How to visualize the sound

In simple terms, sound/music is nothing but a series of signals with a certain frequency and a certain amplitude. We can get the value at each point by using `analogRead()`. But, this function would be much too slow for sampling audio. So, we will use the microcontroller's analog-to-digital converter, which automatically takes repeated analog intervals at precise intervals. We will learn both ways of sampling audio. There are a number of algorithms available for analyzing the sampled audio. But, for fast performance, we will use a FFT (Fast Fourier Transform) algorithm.

What is FFT (fast fourier transform)

Fast Fourier Transform is one of the basic and most important numerical algorithms in the field of signal processing. It converts a signal from its time domain (time domain refers to analysis of a signal with respect to time) to frequency domain. Frequency domain refers to the analysis of a mathematical function, or here a signal with respect to frequency.

You can understand the time domain and frequency domain with the following image:

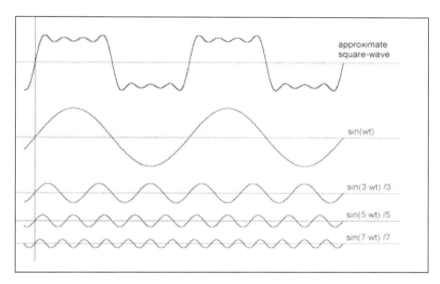

You can decompose any signal into a bunch of sine waves (of different amplitude, frequency, and phase). An analogy for this is the images you see on your computer screen, which are composed of red, green, and blue dots (of different magnitude and frequency).

FFT decomposes sine waves from an arbitrary signal by multiplying the arbitrary signal with a sine wave of a specific frequency. As the match is closer, the resulting summed product value will be higher. Once we have the signal, which is converted to the frequency domain, it becomes easier to analyze.

Sound visualization using Arduino

After understanding the basics of sound visualization, we will move on to implement sound visualization using Arduino. Before we develop our LED Christmas tree, we will develop sound visualization on an LED matrix.

An LED matrix is a combination of 64 LEDs connected together as shown in the following circuit. As Arduino doesn't have 64 pins, we can't connect individual pins to control each LED. Instead, we will use the concept of multiplexing:

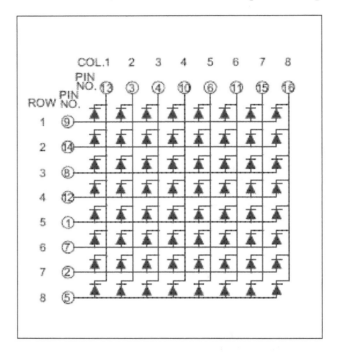

With the use of multiplexing, we can control any number of LEDs with Arduino. For controlling an 8 x 8 LED matrix, we need one multiplexer circuit. We can use the backpack from adafruit for the multiplexer, or a MAX7219 Dot Matrix MCU Control for the multiplexer part:

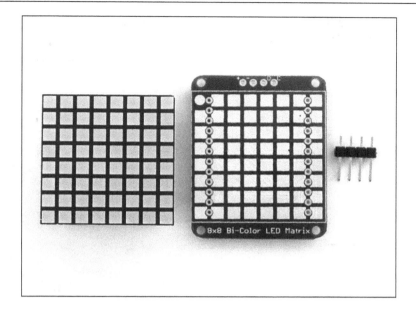

So, now we do not need to use a lot of pins to control this LED matrix. Instead, we require only four pins to control this LED matrix.

You will need the following components for a music-controlled LED matrix:

- Analog Mic Sensor
- 8 x 8 LED matrix with backpack

Connect the mic and LED matrix as shown in the circuit. Connect a 3.3 V pin to the AREF of Arduino and the mic's Vcc in pin. This is important for making a reference of 3.3 V to analog input from the mic. Connect the Arduino 5 V pin to the + pin of the LED matrix. Connect the Arduino analog pin A0 to Mic Output. Connect the Arduino SDA and SCL pin to the D (data) and C (clock) pin of the backpack. Earlier versions of Arduino might not have these pins.

Use analog pins 4 and 5 for this purpose instead. Finally, don't forget to connect Arduino Ground to Mic GND and backpack GND:

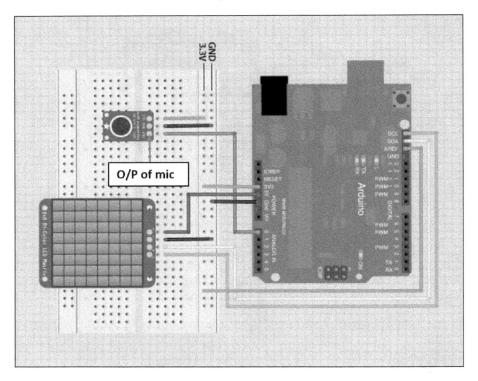

Once you have connected the circuit as shown in the diagram, upload the following code to Arduino:

```
#include <avr/pgmspace.h>
// Useful for Sound Analysis
#include <ffft.h>
#include <math.h>
// Required for communication with I2C device
#include <Wire.h>
#include <Adafruit_GFX.h>
// Required for Adafruit LED backpack
#include <Adafruit_LEDBackpack.h>

// Microphone connects to Analog Pin 0. ADC Pin 0 for Arduino Uno
#define ADC_INPUT 0

int16_t audio_capture[FFT_N];
```

```
complex_t fft_buffer[FFT_N];
uint16_t output_spectrum[FFT_N / 2];
volatile byte buffer_position = 0;

byte peakValue[8],dotCount = 0,colCount = 0;
int col[8][10], minAvgLevel[8], maxAvgLevel[8], colDiv[8];

static const uint8_t PROGMEM
// Noise to be removed from each column. Adjust the values as per the
requirements
noiseToDeduct[64] = { 8, 6, 6, 5, 3, 4, 4, 4, 3, 4, 4, 3, 2, 3, 3, 4,
            2, 1, 2, 1, 3, 2, 3, 2, 1, 2, 3, 1, 2, 3, 4, 4,
            3, 2, 2, 2, 2, 2, 2, 1, 3, 2, 2, 2, 2, 2, 2, 2,
            2, 2, 2, 2, 2, 2, 2, 2, 2, 2, 2, 2, 2, 3, 3, 4
            },
// Equalizer to remove the noise and neutralize the noise at the bass
end.
equalizer[64] = {
  255, 175, 218, 225, 220, 198, 147, 99, 68, 47, 33, 22, 14,  8,  4,
2,
   0,  0,  0,  0,  0,  0,  0,  0,  0,  0,  0,  0,  0,  0,  0,  0,
   0,  0,  0,  0,  0,  0,  0,  0,  0,  0,  0,  0,  0,  0,  0,  0,
   0,  0,  0,  0,  0,  0,  0,  0,  0,  0,  0,  0,  0,  0,  0,  0
},


// We want to fit the output of FFT spectrum to 8 columns.
// All the bins from the output spectrum is not useful.
// we will following bins for the column output.
// Below table contains details about number of bins to take and index
from where it will start
// along with the nin numbers.
column0[] = {  2,  1,
               111,  8
            },
column1[] = {  4,  1,
              19, 186,  38,  2
            },
column2[] = {  5,  2,
              11, 156, 118, 16,  1
            },
column3[] = {  8,  3,
               5, 55, 165, 164, 71, 18,  4,  1
            },
column4[] = { 11,  5,
```

```
                      3,   24,   89, 169, 178, 118,   54,   20,    6,    2,    1
                     },
column5[] = { 17,    7,
                      2,    9,   29,   70, 125, 172, 185, 162, 118,   74,
                     41,   21,   10,    5,    2,    1,    1
                     },
column6[] = { 25,   11,
                      1,    4,   11,   25,   49,   83, 121, 156, 180, 185,
                    174, 149, 118,   87,   60,   40,   25,   16,   10,    6,
                      4,    2,    1,    1,    1
                     },
column7[] = { 37,   16,
                      1,    2,    5,   10,   18,   30,   46,   67,   92, 118,
                    143, 164, 179, 185, 184, 174, 158, 139, 118,   97,
                     77,   60,   45,   34,   25,   18,   13,    9,    7,    5,
                      3,    2,    2,    1,    1,    1,    1
                     },
// This contains list of all the data bin for all 8 columns
* const binsToUse[]  = {
  column0, column1, column2, column3,
  column4, column5, column6, column7
};

Adafruit_BicolorMatrix LEDmatrix = Adafruit_BicolorMatrix();

void setup() {
  uint8_t i, j, nBins, binNum, *outputData;

  memset(peakValue, 0, sizeof(peakValue));
  memset(col , 0, sizeof(col));

  for (i = 0; i < 8; i++) {
    minAvgLevel[i] = 0;
    maxAvgLevel[i] = 512;
    outputData = (uint8_t *)pgm_read_word(&binsToUse[i]);
    nBins = pgm_read_byte(&outputData[0]) + 2;
    binNum = pgm_read_byte(&outputData[1]);
    for (colDiv[i] = 0, j = 2; j < nBins; j++)
      colDiv[i] += pgm_read_byte(&outputData[j]);
  }

  LEDmatrix.begin(0x70);
```

```
  // Init ADC free-run mode; f = ( 16MHz/prescaler ) / 13 cycles/
conversion
  ADMUX  = ADC_INPUT; // Channel sel, right-adj, use AREF pin
  ADCSRA = _BV(ADEN)  | // ADC enable
           _BV(ADSC)  | // ADC start
           _BV(ADATE) | // Auto trigger
           _BV(ADIE)  | // Interrupt enable
           _BV(ADPS2) | _BV(ADPS1) | _BV(ADPS0); // 128:1 / 13 = 9615
Hz
  ADCSRB = 0;
  DIDR0  = 1 << ADC_INPUT;
  TIMSK0 = 0;

  sei(); // Enable interrupts
}

void loop() {
  uint8_t  i, x, L, *outputData, nBins, binNum, weighting, c;
  uint16_t minLvl, maxLvl;
  int      level, y, sum;

  while (ADCSRA & _BV(ADIE)); // Wait for audio sampling to finish

  fft_input(audio_capture, fft_buffer);   // Samples -> complex #s
  buffer_position = 0;                     // Reset sample counter
  ADCSRA |= _BV(ADIE);                // Resume sampling interrupt
  fft_execute(fft_buffer);            // Process complex data
  fft_output(fft_buffer, output_spectrum); // Complex -> spectrum

  // Remove noise and apply equalizers
  for (x = 0; x < FFT_N / 2; x++) {
    L = pgm_read_byte(&noiseToDeduct[x]);
    output_spectrum[x] = (output_spectrum[x] <= L) ? 0 :
                 (((output_spectrum[x] - L) * (256L - pgm_read_
byte(&equalizer[x]))) >> 8);
  }

  // Fill background w/colors, then idle parts of columns will erase
  LEDmatrix.fillRect(0, 0, 8, 3, LED_RED);    // Upper section
  LEDmatrix.fillRect(0, 3, 8, 2, LED_YELLOW); // Mid
  LEDmatrix.fillRect(0, 5, 8, 3, LED_GREEN);  // Lower section

  // Downsample spectrum output to 8 columns:
  for (x = 0; x < 8; x++) {
```

```
    outputData   = (uint8_t *)pgm_read_word(&binsToUse[x]);
    nBins  = pgm_read_byte(&outputData[0]) + 2;
    binNum = pgm_read_byte(&outputData[1]);
    for (sum = 0, i = 2; i < nBins; i++)
      sum += output_spectrum[binNum++] * pgm_read_
byte(&outputData[i]); // Weighted
    col[x][colCount] = sum / colDiv[x];                   // Average
    minLvl = maxLvl = col[x][0];
    for (i = 1; i < 10; i++) { // Get range of prior 10 frames
      if (col[x][i] < minLvl)
      {
        minLvl = col[x][i];
      }
      else if (col[x][i] > maxLvl)
      {
        maxLvl = col[x][i];
      }
    }
    // minLvl and maxLvl indicate the extents of the FFT output, used
    // for dynamically setting the min and max level of the column.

    if ((maxLvl - minLvl) < 8)
    {
      maxLvl = minLvl + 8;
    }
    minAvgLevel[x] = (minAvgLevel[x] * 7 + minLvl) >> 3; // Dampen
min/max levels
    maxAvgLevel[x] = (maxAvgLevel[x] * 7 + maxLvl) >> 3; // (fake
rolling average)

    // Second fixed-point scale based on dynamic min/max levels:
    level = 10L * (col[x][colCount] - minAvgLevel[x]) /
            (long)(maxAvgLevel[x] - minAvgLevel[x]);

    // Clip output and convert to byte:
    if (level < 0L)
    {
      c = 0;
    }
    else if (level > 10)
    {
      c = 10; // Allow dot to go a couple pixels off top
    }
    else
```

```
  {
    c = (uint8_t)level;
  }

  if (c > peakValue[x])
  {
    peakValue[x] = c; // Keep dot on top
  }

  if (peakValue[x] <= 0) // No output
  {
    LEDmatrix.drawLine(x, 0, x, 7, LED_OFF);
    continue;
  }
  else if (c < 8) // Partial column?
  {
    LEDmatrix.drawLine(x, 0, x, 7 - c, LED_OFF);
  }

  // The 'peak' dot color varies, but doesn't necessarily match
  // the three screen regions...yellow has a little extra influence.
  y = 8 - peakValue[x];
  if (y < 2)
  {
    LEDmatrix.drawPixel(x, y, LED_RED);
  }
  else if (y < 6)
  {
    LEDmatrix.drawPixel(x, y, LED_YELLOW);
  }
  else
  {
    LEDmatrix.drawPixel(x, y, LED_GREEN);
  }
}

LEDmatrix.writeDisplay();

// Every third frame, make the peak pixels drop by 1:
if (++dotCount >= 3)
{
  dotCount = 0;
  for (x = 0; x < 8; x++)
  {
```

```
        if (peakValue[x] > 0) peakValue[x]--;
    }
  }

  if (++colCount >= 10)
  {
    colCount = 0;
  }
}

ISR(ADC_vect) { // Audio-sampling interrupt
  static const int16_t noiseThreshold = 4;
  int16_t sample = ADC; // values between 0-1023

  audio_capture[buffer_position] =
    ((sample > (512 - noiseThreshold)) &&
    (sample < (512 + noiseThreshold))) ? 0 :
    sample - 512; // Sign-convert for FFT; -512 to +511

  if (++buffer_position >= FFT_N) ADCSRA &= ~_BV(ADIE); // Turn off
the interrupt once buffer is full
}
```

Let's understand the code in detail. Here, as we discussed earlier, we are using onboard ADC for sampling the audio. While dealing with sampling/signal processing, one of the most useful features of Arduino that comes in handy is interrupts. Before we understand how interrupts work, we need to know about interrupts.

Interrupts allow certain important tasks to happen in the background and will interrupt the execution flow when a particular event occurs. In our case, interrupts are used for sampling the audio. Once the audio is processed with an FFT algorithm and is rendered on the display, it will again sample the audio.

Although we can't hear all the sounds, there is some background noise present around us. We have to remove the ambient noise before processing music to render on the LED matrix. So, we have stored 64 values in the array noiseToDeduct, which will be deducted from the signal afterwards. If you are getting more noise in the output, you can adjust the noise by setting the different values in the array.

Apart from the noise, we also need to equalize the sound towards the bass end of the sound. So, we are storing different values in an array equalizer. After noise is deducted from the output of the filter, we will normalize the sound/output spectrum.

Also, the output spectrum of FFT will be having much more buckets/samples then we can use. The bottom-most and several at the top of the spectrum are either noisy or out of range. So, we will use the FFT spectrum output values stored in the column0, column1, column2, column3, column4, column5, column6, and column7 array. The array contains the number of the output values to be used and the starting index of the bins along with their weights. Although these values can be changed as per our requirements, these values are derived after thorough testing by adafruit.

Two new functions that we have used here are: pgm_read_word and pgm_read_byte. Both functions are provided by the avr/pgmspace.h library. pgm_read_word is used for reading a word from the program space with a 16-bit (near) address. Similarly, pgm_read_byte is used for reading a byte from the program space with a 16-bit (near) address:

```
outputData   = (uint8_t *)pgm_read_word(&binsToUse[x]);
    nBins  = pgm_read_byte(&outputData[0]) + 2;
    binNum = pgm_read_byte(&outputData[1]);
```

For getting data from the binsToUse array, we used pgm_read_word. While reading a particular value from outputData, we have used pgm_read_byte.

Here, we are using a 128-bit buffer for storing the sampled audio. Whenever the buffer is full, it will process the data by passing it to the FFT filter. After the output spectrum from the FFT, the signal is again down-sampled:

```
        sum += output_spectrum[binNum++] * pgm_read_
byte(&outputData[i]); // Weighted
    col[x][colCount] = sum / colDiv[x];                    // Average
```

We are finding a weighted average for each column by selecting the bin from the predefined table initialized at the beginning. After finding the weighted average for eight columns, we are writing these values to the LED matrix display after storing the minimum and maximum values of the output spectrum. These dynamic minimum and maximum values are useful for making the display interesting, even at a low volume.

After understanding sound visualization using FFT, we will now develop an LED Christmas tree that syncs its lighting with beats.

Developing an LED Christmas tree

We are now familiar with the concept of sound visualization. We have also learnt about controlling an LED matrix with music. Now, we will develop an LED Christmas tree, which will blink the LEDs as per the music beats.

To develop the basic circuit which responds to the beats, connect the circuit as shown in the following image:

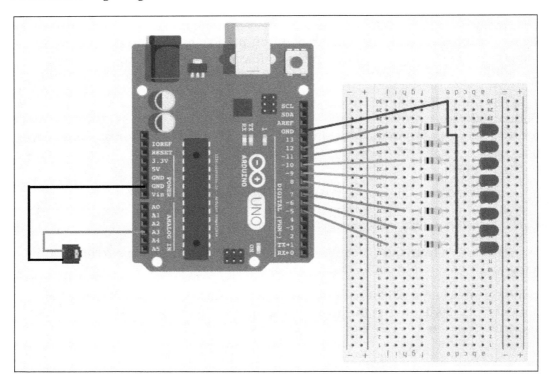

We will connect the audio input/mic to the analog pin 3 of the Arduino. We have connected LEDs to pins 5 to 12.

Once you have connected the circuit as mentioned, upload the following code on the Arduino:

```
#include <fix_fft.h>

int LEDPins[] = {5, 6, 7, 8, 9, 10, 11, 12};
int x = 0;
char imaginary[128], inputSignal[128];
char outputAverage[14];
int i = 0, inputValue;
#define AUDIOPIN 1

void setup()
{
  for (int i = 0; i < 8; i++)
```

```
  {
    pinMode(LEDPins[i], OUTPUT);
  }
  Serial.begin(9600);
}

void loop()
{
  for (i = 0; i < 128; i++) {
    inputValue = analogRead(AUDIOPIN);
    inputSignal[i] = inputValue;
    imaginary[i] = 0;
  };
  fix_fft(inputSignal, imaginary, 7, 0);
  for (i = 0; i < 64; i++) {
    inputSignal[i] = sqrt(inputSignal[i] * inputSignal[i] +
imaginary[i] * imaginary[i]);  // this gets the absolute value of the
values in the
    //array, so we're only dealing with positive numbers
  };

  // average bars together
  for (i = 0; i < 14; i++) {
    outputAverage[i] = inputSignal[i * 4] + inputSignal[i * 4 + 1] +
inputSignal[i * 4 + 2] + inputSignal[i * 4 + 3]; // average together
    outputAverage[i] = map(outputAverage[i], 0, 30, 0, 9);
  }
  int value = outputAverage[0];//0 for bass
  writetoLED(value);
}

void writetoLED(int mappedSignal)
{
  if (mappedSignal > 8)
  {
    for (int i = 0; i < 8; i++)
    {
      digitalWrite(LEDPins[i], HIGH);
    }
  }
  else if (mappedSignal > 7)
  {
    for (int i = 0; i < 7; i++)
    {
```

```
      digitalWrite(LEDPins[i], HIGH);
    }
    for (int i = 7; i < 8; i++)
    {
      digitalWrite(LEDPins[i], LOW);
    }
  }
  else if (mappedSignal > 6)
  {
    for (int i = 0; i < 6; i++)
    {
      digitalWrite(LEDPins[i], HIGH);
    }
    for (int i = 6; i < 8; i++)
    {
      digitalWrite(LEDPins[i], LOW);
    }
  }
  else if (mappedSignal > 5)
  {
    for (int i = 0; i < 5; i++)
    {
      digitalWrite(LEDPins[i], HIGH);
    }
    for (int i = 5; i < 8; i++)
    {
      digitalWrite(LEDPins[i], LOW);
    }
  }
  else if (mappedSignal > 4)
  {
    for (int i = 0; i < 4; i++)
    {
      digitalWrite(LEDPins[i], HIGH);
    }
    for (int i = 4; i < 8; i++)
    {
      digitalWrite(LEDPins[i], LOW);
    }
  }
  else if (mappedSignal > 3)
  {
    for (int i = 0; i < 3; i++)
    {
```

```
        digitalWrite(LEDPins[i], HIGH);
      }
      for (int i = 3; i < 8; i++)
      {
        digitalWrite(LEDPins[i], LOW);
      }
    }
    else if (mappedSignal > 2)
    {
      for (int i = 0; i < 2; i++)
      {
        digitalWrite(LEDPins[i], HIGH);
      }
      for (int i = 2; i < 8; i++)
      {
        digitalWrite(LEDPins[i], LOW);
      }
    }
    else if (mappedSignal > 1)
    {
      for (int i = 0; i < 1; i++)
      {
        digitalWrite(LEDPins[i], HIGH);
      }
      for (int i = 1; i < 8; i++)
      {
        digitalWrite(LEDPins[i], LOW);
      }
    }
    else
    {
      for (int i = 0; i < 8; i++)
      {
        digitalWrite(LEDPins[i], LOW);
      }
    }
  }
}
```

Here, we are directly reading the signal from the mic by using `analogRead()`. We are taking 128 readings. After taking all the readings, we are processing those readings using the `fix_fft.h` library, which is available at `http://forum.arduino.cc/index.php/topic,38153.0.html`. After storing both `fix_fft.h` and `fix_fft.cpp` in your machine, you need to import this library in your Arduino code.

After getting the output spectrum from the FFT, we are taking the average value of the output for turning on or off the LEDs based on the average value. As the bass part most often provides harmonics and rhythmic support, we are using the bass value for controlling the LEDs. We are getting the bass value as `outputAverage[0]`. Mapping of the average value from 0 to 9 helps in setting the LEDs to on or off easily.

For artistic purposes, we will connect multiple LEDs around the Christmas tree model. So, the Christmas tree will sync its lighting with the beats of the music.

Summary

In this chapter, we started with the basics of sound visualization. After understanding the FFT algorithm, we move on to visualize the sound with Arduino using the same on an 8 x 8 dot matrix LED display. At the end of this chapter, we developed an LED Christmas tree, which syncs its light as per the beats in the music. In the next chapter, we will move on to develop "persistence of vision".

6
Persistence of Vision

So far in this book, we have made all things that are stationary in nature — that is, they can't move. In the final project of this book, we will create an even more intensive experience by moving LEDs using motors. We will create a Persistence of Vision wand using an LED array and motor. But, first of all, you will get introduced to LED arrays and motors. Along with the different type of motors, you will get to know about their pros and cons. In this chapter, we will cover the following topics:

- Persistence of Vision
- Programming an LED array
- Controlling a motor using Arduino
- Synchronizing LED array timing based on the speed of the motor

Creating your own Persistence of Vision

One of the five sensory organs of our body, the eye is a remarkable instrument that helps us to process light in such a way that our mind can create meaning from it. Persistence of Vision refers to an optical illusion where multiple different images blend into a single image in the human mind.

The Persistence of Vision illusion plays a role in keeping the world from going pitch black every time we blink our eyes. Whenever a retina is hit by light, it keeps an impression of the light for about a tenth of a second after the light source is removed. Due to this, the eye can't distinguish between changes that occur faster than this retention period. This similar phenomenon is used in motion pictures or, as we call it, "flicks". The motion picture creates an illusion by rapidly sequencing individual photographs. Usually for motion pictures, the rate of frames per second is 24, which leads to a flicker-free picture. If frames per second is kept below 16, the mind can distinguish between the images, which leads to flashing images, or flicker.

Look at the following image; when you move your hands back and forth as demonstrated, you will see a flicker at all the positions:

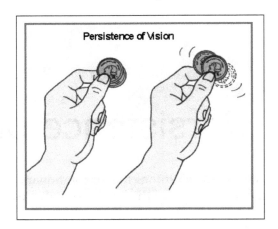

A renowned professor at the University of Central Arkansas quoted this:

> *"The notion of 'persistence of vision' seems to have been appropriated from psychology in the first decade of the century, the period during which cinema came into being. But while most film scholars accepted 'persistence of vision' as the perceptual basis of the medium and proceeded to theorise about the nature, meaning and functioning of cinema from that base, perceptual psychologists continued to question the mechanisms involved in motion perception; and they have achieved insights that necessitate the re-thinking of many conclusions reached by film scholars during the past 50 years."*

After getting introduced to the concept of Persistence of Vision, let's dive into how we can make our own PoV. For that we will need the following components:

- Arduino
- LED array
- DC motor
- Resistor
- L293D motor driver
- Wooden material for making the base of the PoV

In the upcoming section, you will learn how to use these components to create your own Persistence of Vision.

Programming an LED array

An LED array is nothing but a few LEDs connected together. Usually, an LED array comes in sizes of eight LEDs and 16 LEDs. You can control an LED array directly using the digitalWrite() function. Apart from using the digitalWrite() function, you can control LEDs directly using port-level communication. On Arduino, we have three ports: ports B, C, and D:

- Port B: Digital pins 8 to 13
- Port C: Analog pins
- Port D: Digital pins 0 to 7

Each port is controlled by three DDR registers. These registers are defined variables in Arduino as DDRB, DDRC, and DDRD. Using these variables, we can make the pins either as input or output in the setup function.

You can use the following syntax to initialize the pins:

```
DDRy = Bxxxxxxxx
```

Here, y is the name of the port (B/C/D) and x is the value of the pin that determines if the pin is input or output. We will use 0 for input and 1 for output. LSB (least significant byte) is the lowest pin number for that register.

To control pins using this port manipulation, we can use the following syntax:

```
PORTy = Bxxxxxxxx
```

Here, y is the name of the port (B/C/D), and make x equal 1 for making a pin HIGH and 0 for making the pin LOW.

You can use the following code to control the LEDs using port level communication:

```
void setup()
{
  DDRD = B11111111; // set PORTD (digital 7-0) to outputs
}

void loop()
{
  PORTD = B11110000; // digital 4~7 HIGH, digital 3-0 LOW
  delay(2000);
  PORTD = B00001111; // digital 4~7 LOW, digital 3-0 HIGH
  delay(2000);
}
```

There are a few pros and cons for using a port manipulation technique. Following are the disadvantages of using a port manipulation technique:

- The code becomes more difficult to debug and maintain and it takes a lot of time to understand the code.
- The code becomes less portable. If you use `digitalWrite()` and `digitalRead()`, it is much easier to write a code that will run on all microcontrollers, whereas ports and registers can be different for each kind of microcontroller.
- You might cause unintentional malfunctions with direct port access, as pin 0 is the receive line for the serial port. If you accidently make it input, it may cause your serial port to stop working.

There are a few advantages that a port manipulation technique has over normal code practices:

- In case of time constraint uses of Arduino, you would need to turn pins on or off very quickly. Using direct port access, you can save many machine cycles.
- Also, if you are running low on program memory, you can use these techniques to make your code smaller.

Different types of motors

Depending upon your project needs, you can choose from the variety of motors available in the market. For hobby electronics, you will mostly use either DC motor, servo motor, or stepper motor. The following image shows all three types of motors. From left to right, DC motor, servo motor, and stepper motor:

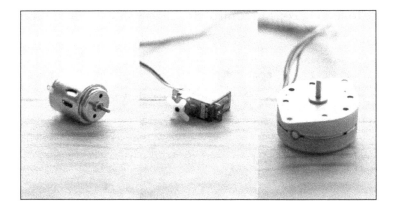

Let's get an overview about all the different types of motors.

DC motors

DC (Direct Current) motors are two-wire continuous rotational motors. When power is supplied to the motor, it will start running and will stop once power is removed. Most DC motors run at high speed, that is, high RPM (rotations per minute). The speed of the DC motor is controlled using the PWM technique (as discussed in *Chapter 2, Project 1 – LED Night Lamp*). The duty cycle will determine the speed of the motor. The motor seems to be continuously running as each pulse is very rapid.

Servo motors

Servo motors are self-contained electric devices that rotate or push parts of a machine with great precision. A servo motor uses closed loop position feedback to control its motion and final position. Servo motors are intended for use in more specific tasks, where position needs to be accurate, such as moving a robotic arm. For this purpose, servo motors are an assembly of a DC motor, control circuit, potentiometer, and a gearing set.

The angle of rotation is limited to 180 degrees compared to the free run of a normal DC motor. They usually have three wires consisting of power, ground, and signal. To run the servo motors, continuous power is required. By giving a signal of proper value to the signal pin of the servo, one can control the position of the servo motor.

When a servo is given the signal to move, if any external force is applied to change its position, the servo motor will try to hold on to its position. A servo motor uses an integrated controller circuit to position itself.

Stepper motors

A stepper motor is a DC motor that moves in discrete steps. A stepper motor utilizes multiple toothed electromagnets arranged around a central gear to set the position. A stepper motors require an external control circuit to energize each electromagnet and make the motor shaft turn.

Stepper motors are available in two types, that is, unipolar and bipolar. Bipolar motors are the strongest type of stepper motor, having four or eight leads. Unipolar stepper motors are simpler compared to bipolar motors. Unipolar motors can step without reversing the direction of the current in the coils, because it is having a centre tap internally. However, because of the centre tap, unipolar motors have less torque compared to bipolar motors.

The basic difference between servo motors and stepper motors is the type of motor and how it is controlled. Stepper motors typically use 50 to 100 pole brushless motors, while servo motors have only 4 to 12 poles.

Different applications of motors

This is a very condensed overview of a somewhat complicated field:

- **DC motors**: Used in fans, car wheels, and so on, which need fast and continuous rotation motors.

- **Servo motors**: Servo motors are usually suited for robotic arms/legs where fast, high torque, and accurate rotation within a limited angle is required.

- **Stepper motors**: For devices where precise rotation and accurate control is required, stepper motors are used. Stepper motors have an advantage over servo motors in positional control because a stepper motor has positional control due to its nature of rotation by fractional increments.

Controlling a DC motor using Arduino

In this chapter, we will get to know how to control a DC motor with Arduino.

You can also run the DC motor by using the same code as the LED blink. We can consider the motor as an LED. As discussed earlier, a DC motor is a two-wired motor. One wire is the positive supply and other is ground. If we connect the positive terminal of the motor to the positive terminal of the battery, and the negative terminal of the motor to the negative terminal of the battery, the motor will run in one direction, and if we reverse the connection, the motor will run in the reverse direction.

By connecting the motor to two digital pins of the Arduino, we can control the direction of the motor. In the following basic code, we will run the motor in one direction for five seconds and then we will reverse the direction of the motor. Connect pin 3 and pin 4 of the Arduino with the two wires of the motor:

```
int motorPos = 3;
int motorNeg = 4;

void setup() {
  pinMode(motorPos, OUTPUT);
  pinMode(motorNeg, OUTPUT);
}

void loop() {
  //run the motor in one direction
  digitalWrite(motorPos, HIGH);
  digitalWrite(motorNeg, LOW);
  delay(5000); //Run for 5 seconds
  // Reverse the direction of the motor
```

```
    digitalWrite(motorPos, LOW);
    digitalWrite(motorNeg, HIGH);
    delay(5000);
}
```

In the preceding code, we are giving opposite outputs to both pins 3 and 4, which is useful in defining the direction of the motor.

Although this method of controlling DC motors directly with Arduino seems very easy, it has its own disadvantages. From the Arduino I/O pin, one can draw the maximum current of 20 mA. So, if we connect the heave load to Arduino, Arduino might get busted. For this purpose, we use an L293D chip, which is compatible with H-bridge connections. H-bridge is a circuit, which can drive the motor in both directions. Before we connect L293D to Arduino, check out all the pin details in the following image:

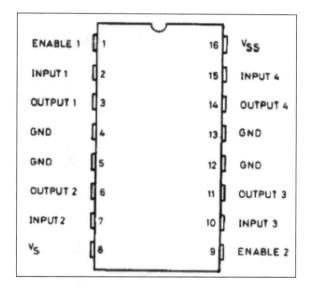

Connect **Enable1** and **Enable2** with a 5 V constant logic supply. L293D IC is designed in such a way that the left pins of the IC can be used to control one motor and the right pins of the IC can be used to control another motor in both directions. For controlling one motor, give input to pin 2 and 7, which will give output at pin 3 and pin 7. Pin 8 is the power supply for the motors. Connect a 5 V logic supply to pin 16.

In most cases, a voltage regulator is required as the controller can't handle voltage more than 5 V. But, Arduino UNO has an in-built voltage regulator. Although you can give up to 20 V to the Arduino UNO, it is recommended to give input voltage up to 12 V.

As you can see in the preceding image, two motors can be controlled with one single chip, with normal voltage (9 V).

We will connect Arduino UNO as a controller and will connect inputs at pins 3 and 4. We will connect one motor at pins 3 and 4, and the other motor at pins 7 and 8. As shown in the following image, we can make a simple robot by connecting a caster as a third wheel after fitting those motors to the chassis:

We will make this robot do a simple repetitive task using this L293D and Arduino. Connect the Arduino control signal/output to the motors to pins 2, 7, 9, and 15 of the L293D.

Connect the motor between pins 3 and 6 of L293D. Once you have made the same connection for the other side of L293D, that is, pins 9 to 15, your circuit will look like the following image:

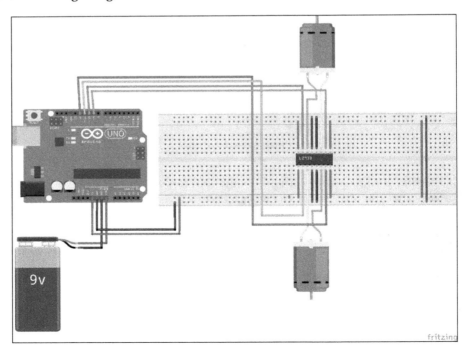

After checking all the connections once again, upload the following code to Arduino:

```
int leftMotorPos = 10;
int leftMotorNeg = 9;
int rightMotorPos = 12;
int rightMotorNeg = 13;

void setup()
{
  pinMode(leftMotorPos, OUTPUT);
  pinMode(rightMotorPos, OUTPUT);
  pinMode(leftMotorNeg, OUTPUT);
  pinMode(rightMotorNeg, OUTPUT);
}

void loop()
{
  forward();
  delay(5000);
  right();
  delay(5000);
  left();
  delay(5000);
  reverse();
  delay(5000);
  stopAll();
  delay(5000);
}

void forward() {
  digitalWrite(rightMotorPos, HIGH);
  digitalWrite(leftMotorPos, HIGH);
  digitalWrite(rightMotorNeg, LOW);
  digitalWrite(leftMotorNeg, LOW);
}

void left() {
  digitalWrite(rightMotorPos, HIGH);
  digitalWrite(leftMotorPos, LOW);
  digitalWrite(rightMotorNeg, LOW);
  digitalWrite(leftMotorNeg, LOW);
}

void right() {
```

```
    digitalWrite(rightMotorPos, LOW);
    digitalWrite(leftMotorPos, HIGH);
    digitalWrite(rightMotorNeg, LOW);
    digitalWrite(leftMotorNeg, LOW);
}

void reverse() {
    digitalWrite(rightMotorPos, LOW);
    digitalWrite(leftMotorPos, LOW);
    digitalWrite(rightMotorNeg, HIGH);
    digitalWrite(leftMotorNeg, HIGH);
}

void stopAll() {
    digitalWrite(rightMotorNeg, LOW);
    digitalWrite(leftMotorNeg, LOW);
    digitalWrite(rightMotorPos, LOW);
    digitalWrite(leftMotorPos, LOW);
}
```

As per the preceding circuit and code, we are not giving input directly to the motors; rather, we are giving inputs to the L293D, which in turn will provide sufficient power to run the motors.

 Make sure to connect the grounds between Arduino and L293D. Otherwise, the grounds will be in floating mode and the motor will run abruptly, that is, the motor might run sometimes and might not run sometimes.

Another way to control the motor using L293 is to give a signal to an enable pin. By controlling enable pins' input, you can control the motors. Here, we are giving control to the motor input and the enable pins are given high input continuously.

Synchronizing an LED array with a motor

In the previous sections of this chapter, we learned about controlling an LED array and DC motor using Arduino:

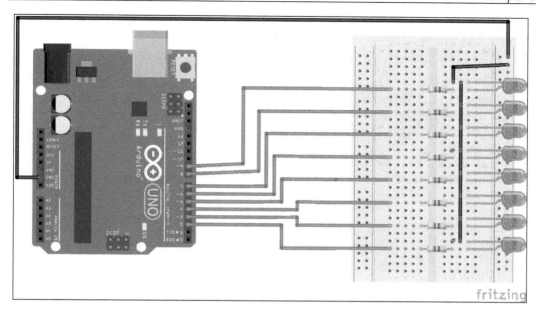

Once you have connected the circuit as shown in the preceding image, upload the following code to Arduino. In the following code, we are writing the "Hello world" of persistence of vision:

```
int LEDPins[] = {2, 3, 4, 5, 6, 7, 8, 9};
int noOfLEDs = 8;

//data corresponding to the each alphabet and a few characters to be
displayed
byte H[] = {B11111111, B11111111, B00011000, B00011000, B00011000,
B00011000, B11111111, B11111111};
byte E[] = {B00000000, B11111111, B11011011, B11011011, B11011011,
B11011011, B11000011, B11000011};
byte L[] = {B00000000, B11111111, B11111111, B00000011, B00000011,
B00000011, B00000011, B00000011};
byte O[] = {B00000000, B11111111, B11111111, B11000011, B11000011,
B11000011, B11111111, B11111111};
byte fullstop[] = {B00000000, B00000000, B00000000, B00000011,
B00000011, B00000000, B00000000, B00000000};
byte comma[] = {B00000000, B00000000, B00000000, B00000110, B00000101,
B00000000, B00000000, B00000000};

// Customize parameters based on the need
int timeBetweenColumn = 2.2;
int timeBtwnFrame = 20;
```

```
int frame_len = 8;

void setup()
{
  int i;
  pinMode(12, OUTPUT);
  pinMode(13, OUTPUT);
  pinMode(11, OUTPUT);
  pinMode(10, INPUT);
  for (i = 0; i < noOfLEDs; i++) {
    pinMode(LEDPins[i], OUTPUT);    // set each pin as an output
  }
}

void loop()
{
  int b = 0;
  digitalWrite(12, HIGH);
  digitalWrite(13, HIGH);
  digitalWrite(11, HIGH);
  delay(timeBtwnFrame);
  show(H);
  delay(timeBtwnFrame);
  show(E);
  delay(timeBtwnFrame);
  show(L);
  delay(timeBtwnFrame);
  show(L);
  delay(timeBtwnFrame);
  show(O);
  delay(timeBtwnFrame);
}

void show( byte* image )
{
  int a, b, c;

  // go through all data for all columns in each frame.
  for (b = 0; b < frame_len; b++)
  {
    for (c = 0; c < noOfLEDs; c++)
    {
      digitalWrite(LEDPins[c], bitRead(image[b], c));
    }
```

```
        delay(timeBetweenColumn);
    }
    for (c = 0; c < noOfLEDs; c++)
    {
        digitalWrite(LEDPins[c], LOW);
    }
}
```

Initially, we are setting pins which need to be turned on, while writing any letters. Here, we have written code for letters H, E, L, O, period, and comma.

Here, we are using eight LEDs for displaying PoV. Make sure to connect a resistor between the LED and Arduino. As we have to make the LEDs glow based on the speed of the motor, we can control that by changing the value of `timeBetweenColumn` and `timeBtwnFrame` variables in the code.

By changing the values of these two variables, you should be able to sync the LEDs with the motor speed. One more thing that you can do is initialize the variable at a fixed value, and by using serial communication, change the value of these variables. Using the SoftwareSerial library, you can easily accomplish this.

Bringing your efforts to life

Once you know how to control motors and LEDs using Arduino, the final step is to put everything that you have learned so far and make it a standalone product. There are multiple ways you can achieve this:

- The simplest method is to use your hands.
- Using two different Arduinos or using external motors
- Using existing real-life devices

Using your hands for rotation

Even though you learned about controlling motors and LEDs, if you are doing this for the first time, you will take some time to understand the synchronization of LEDs and motors. The easiest way to test Persistence of Vision is to use your hands for rotation. Before you start rotating the complete setup, make sure you have uploaded the latest sketch to Arduino, connected some standalone battery/power source, and firmly fixed your LEDs on some base. As a base, you can use a thermocol sheet or cardboard, and fix them using some electrical tape.

Once you have fixed all the materials, you can rotate your PoV, and you can see your efforts coming to life, as shown in the following image:

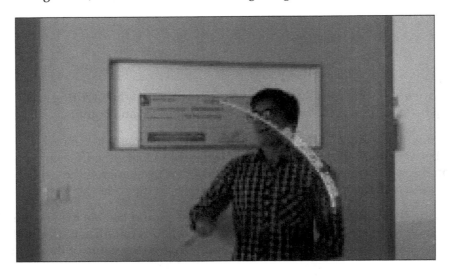

Using two different Arduinos or external motors

Once you have fixed your Arduino, LEDs, and external battery/power source, you can use one more Arduino and connect motors for rotating the complete structure. At first it seems you can control the motors and LEDs using the same Arduino, however, if you think a little bit more, you will understand why you should not do this. The reason is because of the rotation of the LED structure and the connecting wire between motor and Arduino. If you have an external motor, you can use that and connect the Arduino-LED structure to it.

Use existing real-life devices

Another interesting idea is to use existing real-life devices that you have at your home. Of course, there are many devices which you can use; we have tested it with two devices: a bicycle wheel and a table fan, as shown in the following image. There is no difference in setup compared to the other two methods. In fact, you can use the same setup/structure as in method 1:

Summary

In this final project of this book, you learned about motors and how to control an LED array. After understanding Persistence of Vision, we developed a Persistence of Vision display. After developing all the projects in this book, you should be able to play with different types of LED and make some innovative thing out of them. In the last chapter, we will learn about some of the problems that you might face while developing the projects explained in this book.

7
Troubleshooting and Advanced Resources

In this book, you got introduced to Arduino Pi and its capability, you developed your LED night lamp, remote controlled TV backlight, LED cube, sound visualization, and finally, persistence of vision. There might have been instances when you wanted to know more about certain topics or you were stuck in between. This chapter answers all those questions. In the first section, common troubleshooting techniques are mentioned. The second and last part of the chapter has resources that will be useful if you want to do advanced stuff with Arduino:

- Troubleshooting
- Resources – advanced users

Troubleshooting

This section has answers to some of the common problems that you might face while working with Arduino.

Can't upload program

Assign the correct serial port: in the Arduino Environment program, go to **Tools | Serial Port**, and select the correct serial port. To see what serial port the board is using, connect the board to your computer with the USB cable. From the Windows desktop, right-click on My Computer, then **Properties | Device Manager | Ports (COM & LPT)**. There will be an entry like **USB Serial Port (COM13)** or **Arduino UNO (COM13)**.

This means serial communication port 13 is the one in use:

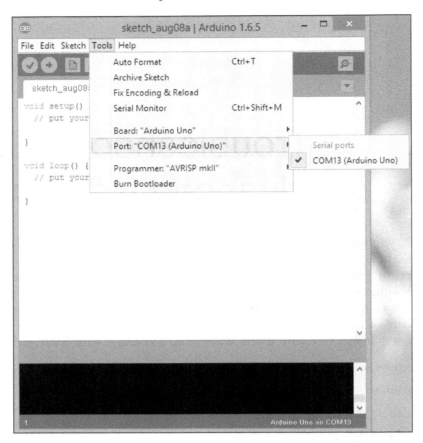

 On upload, you may get error messages like, "Serial port 'COM13' already in use". Try quitting any programs that may be using it.

One of the possible reasons could be that you are running multiple Arduino IDEs on a single machine. Try closing all Arduino IDEs and open the sketch that you want to upload to your Arduino board. Most of the time it will solve the issue. Even after re-opening the Arduino IDE, if you get the same error message, unplug your Arduino and plug it back in. This should solve your issue. In some cases, if you get the same error message again, restart your computer.

LED is dim

This is the most common mistake that beginners make. Sometimes, an LED connected to an Arduino pin is dim. This is because the Arduino pin connected to the LED is not declared as OUTPUT and is not getting the full power from the Arduino board. If you declare the Arduino LED pin as OUTPUT, it will solve the issue and the LED will glow properly.

You can also refer to the troubleshooting section on the Arduino website. You can find solutions to other problems at `https://www.arduino.cc/en/Guide/Troubleshooting`.

Resources – advanced users

This section contains some advanced projects that I think are interesting and fun to build. This last section has some handy and useful resources to take your Arduino journey to the next level.

Projects

Based on all the basic skills of Arduino, LED, and sensor programming that you have, we believe that the following are four projects that might be interesting for you to build.

Twitter Mood Light

This is one of the cool projects that I have seen. It is a way to get a glimpse of the collective human consciousness. It is a way to be alerted with the world's events as they unfold, or when something big happens. Arduino connects directly to any wireless network via the WiFly module. It then searches Twitter for tweets with emotional content and collates the tweets for each emotion. It also does some math, such that the color of the LED fades to reflect the current world mood. Here are a few examples:

- Red for anger
- Yellow for happiness
- Pink for love
- White for fear
- Green for envy
- Orange for surprise
- Blue for sadness

In this project, after getting tweets from the twitter handle for the user, by using the sentiment extraction method, you can get to know about the emotion/mood of the world.

Read more at: http://www.instructables.com/id/Twitter-Mood-Light-The-Worlds-Mood-in-a-Box/.

Secret knock detecting door-lock

You can now keep your secret hideout hidden from intruders with a lock that will only open when it hears the secret knock. This wasn't accepted completely at the beginning, but turned out to be surprisingly accurate at judging knocks. If the precision is turned all the way up it can even tell people apart, even if they give the same knock!

Read more at http://www.instructables.com/id/Secret-Knock-Detecting-Door-Lock/.

LED biking jacket

This is one of the best projects to show your making skills to your friend and the outside world. This project shows you how to build a jacket with turn signals that will let people know where you're headed when you're on your bike. It uses conductive thread and sewable electronics, so your jacket will be soft, wearable, and washable when you're done.

In this project, you will learn to use another type of Arduino which is designed specifically to be wearable—LilyPad.

Read more at http://www.instructables.com/id/turn-signal-biking-jacket/.

Twitter-enabled coffee pot

Tweet-a-pot is the next wave in fancy twitter-enabled devices. This coffee pot enables you to make a pot of coffee from anywhere that has a cell phone reception, using Twitter and an Arduino board. The tweet-a-pot is the easy implementation for remote device control; using a bit of code and some hardware, you can have your very own Twitter-enabled coffee pot.

Read more at: http://www.instructables.com/id/Tweet-a-Pot-Twitter-Enabled-Coffee-Pot/.

Useful resources

In this book, you have learned about Arduino, LEDs, and sensors. If you look at the world of the maker movement and Arduino, we have only scratched the surface, and there are so many things that you need to learn. Here are some resources that we think are useful in taking your skills to the next level.

Hackaday

This is an excellent resource for all sorts of technological wonders. It has lots of Arduino-related projects and easy to follow guides for most of the projects. However, this website is not limited to just Arduino; it has various other resources for almost all DIY technologies. It contains an excellent collection of posts and information to fuel the imagination.

Refer to `http://hackaday.com/`.

The Arduino blog

This is a great resource for all Arduino-related news. It features all the latest Arduino-related hardware, as well as software projects. It is also one of the best places to keep yourself updated with the work that the Arduino team has been doing.

Refer to `https://blog.arduino.cc/`.

The Make magazine

This is a hobbyist magazine that celebrates all kinds of technology. Its blog covers all kinds of interesting do-it-yourself (DIY) technology and projects for inspiration. You can find useful Arduino resources/projects under the "Arduino" section of the website.

Refer to `http://blog.makezine.com/`.

Bildr

Bildr is an excellent resource that provides in-depth, community-published tutorials. As well as providing clear tutorials, Bildr also has excellent illustrations, making the connections easy to follow. Many of the tutorials are Arduino-based and provide all the code and information on the components that you will need.

Refer to `http://bildr.org/`.

Instructables

This is a web-based documentation platform that allows people to share their projects with step-by-step instructions on how to make them. Instructables isn't just about Arduino or even technology, so you can find a whole world of interesting material there.

Refer to `http://www.instructables.com/`.

Tronixstuff

John Boxall's website is a great resource for learning about Arduino. He has dozens of different Arduino projects and demos on his site.

Refer to `http://tronixstuff.com/tutorials/`.

Adafruit

Adafruit is an online shop, repository, and forum for all kinds of kits to help you make your projects work. It is probably one of the best online resources for learning about Arduino and checking out some cool projects.

Refer to `https://learn.adafruit.com/`.

All About Circuits

If you are interested in learning more about electronics and circuit design, this might be the best place for you to learn.

Refer to `http://www.allaboutcircuits.com/`.

Hackerspaces

Hackerspaces are physical spaces where artists, designers, makers, hackers, coders, engineers, or anyone else, can meet to learn, socialize, and collaborate on projects. If you are looking for inspiration and want to meet some awesome people doing some amazing work, find a hackerspace nearby your area and learn from the masters.

Refer to `http://hackerspaces.org/`.

The Arduino forum

This is a great place to get answers to specific Arduino questions. You often find that other people are working through the same problems that you are, so with some thorough searching, you're likely to find the answer to almost any problem.

Refer to `http://arduino.cc/forum/`.

Summary

This chapter provided solutions for some of the common problems that you might face while working with Arduino. The last section of the chapter mentioned a few DIY projects that you might want to pursue with the resources provided.

If you face any issue in any of the projects mentioned in the book, or you notice any typos/errors in any of the chapters, feel free to mail us at `Samarth@outlook.com` and/or `Utsav_shah01@outlook.com` with the subject as the title of the book.

Index

C

connection
  verifying 10

D

DC motor
  about 95
  applications 96
  controlling, Arduino used 96-100

F

fix_fft.h library
  reference 89

H

hack-a-day
  about 111
  URL 111
hackerspaces
  about 112
  URL 112
Hello World program 11
high power LED 16

I

Instructables
  about 112
  URL 112
IR LED
  about 37, 38
  applications 38
IRRemote library
  reference 43
IR sensor
  about 38
  data, receiving from TV remote 41-45
  programming 39, 41
  working mechanism 38, 39

L

LED
  about 16
  application-specific LED 17

high power LED 16
  miniature LED 16
LED array
  programming 93, 94
  synchronizing, with motor 100-103
LED biking-jacket project
  about 110
  reference 110
LED Christmas tree
  developing 85-90
LED cube
  board, fixing 65-68
  construction 61-64
  designing 59
  mistakes, to avoid 64
  principle, behind design 60, 61
  required components 59
LED fading
  about 23
  PWM, using on Arduino 24, 25
LED night lamp
  developing 31
LED strip
  about 45, 46
  controlling, with Arduino 46-55
Light Dependent Resistor (LDR) 32

M

Make magazine
  about 111
  URL 111
miniature LED 16
mood lamp, creating
  about 25
  designing 27-31
  LED night lamp, developing 31
  maintained switch 31
  momentary switch 31
  Pixar lamp 32-34
  RGB LEDs color, changing 26, 27
  RGB LED, using 26
  switch 31
motors
  applications 96
  different types 94
  LED array, synchronizing with 100-103

www.ingramcontent.com/pod-product-compliance
Lightning Source LLC
Chambersburg PA
CBHW060151060326
40690CB00018B/4069